RAPIDLY QUENCHED METALS

A Bibliography, 1973–1979

IFI DATA BASE LIBRARY

COMPUTER TECHNOLOGY
Logic, Memory, and Microprocessors – A Bibliography
A. H. Agajanian

ACOUSTIC EMISSION
A Bibliography with Abstracts
Thomas F. Drouillard

MICROELECTRONIC PACKAGING
A Bibliography
A. H. Agajanian

SPEECH COMMUNICATION AND THEATER ARTS
A Classified Bibliography of Theses and Dissertations, 1973–1978
Merilyn D. Merenda and James W. Polichak

ATOMIC GAS LASER TRANSITION DATA
A Critical Evaluation
William Ralph Bennett, Jr.

BEHAVIORAL DEVELOPMENT OF NONHUMAN PRIMATES
An Abstracted Bibliography
Faren R. Akins, Gillian S. Mace, John W. Hubbard, and Dianna L. Akins

MASTER TABLES FOR ELECTROMAGNETIC DEPTH SOUNDING INTERPRETATION
Rajni K. Verma

SPECTROSCOPIC REFERENCES TO POLYATOMIC MOLECULES
V. N. Verma

RAPIDLY QUENCHED METALS
A Bibliography, 1973–1979
C. Suryanarayana

MOSFET TECHNOLOGIES
A Comprehensive Bibliography
A. H. Agajanian

RAPIDLY QUENCHED METALS

A Bibliography, 1973–1979

Compiled by

C. Suryanarayana

Department of Metallurgical Engineering
Banaras Hindu University
Varanasi, India

IFI/PLENUM • NEW YORK-WASHINGTON-LONDON

Library of Congress Cataloging in Publication Data

Suryanarayana, C

Rapidly quenched metals, a bibliography, 1973–1979.

(IFI data base library)

Includes indexes.

1. Metals–Quenching–Bibliography. 2. Metals–Heat treatment–Bibliography.
I. Title.

Z6679.H4S96 [TN752.Q4] 016.6713'6 80-21936

ISBN 978-1-4684-9136-4 ISBN 978-1-4684-9134-0 (eBook)
DOI 10.1007/978-1-4684-9134-0

A Division of Plenum Publishing Corporation
227 West 17th Street, New York, N.Y. 10011

Softcover reprint of the hardcover 1st edition 1980

ACKNOWLEDGEMENTS

During the compilation of this Bibliography, I had been very fortunate in getting assistance from several friends and well-wishers. I am most grateful to Professor Tsuyoshi Masumoto of the Research Institute for Iron, Steel and Other Metals, Tohoku University, Sendai, Japan and Professor R.W.Cahn of the University of Sussex, England, who have continuously encouraged me in undertaking this Project and given valuable advice from time to time. Professor Pol Duwez of the California Institute of Technology, Pasadena, USA and Professor T.R.Anantharaman of Banaras Hindu University, Varanasi, India have been a source of inspiration.

A work of this nature involves co-operation of several persons from different laboratories. I am happy to state that I had the benefit of advice and the pleasure of receivng lists of publications from people actively engaged in this field of research - too numerous to list. I am most thankful to all of them.

Miss Hiromi Kusaka has put in lot of effort to prepare the manuscript in a cameraready form and I am appreciative of her work.

A major portion of this work was completed during my stay at the Tohoku University in Sendai, Japan under a research program sponsored by the Japan Society for the Promotion of Science. I wish to thank them for giving me an opportunity to come to Japan and also for financial assistance.

May 30,1980 C.Suryanarayana

INTRODUCTION

Professor Pol Duwez of the California Institute of Technology, Pasadena, USA, developed a method in 1960 to solidify metallic melts at cooling rates exceeding about 10^5 K/sec. It was shown that under these non-equilibrium conditions it was possible to obtain supersaturated solid solutions, metastable crystalline intermediate phases and even amorphous phases under appropriate quenching conditions and for suitable alloy compositions. Subsequently, several other techniques have also been developed to achieve these non-equilibrium effects. A common feature of all these rapid quenching techniques (now collectively referred to as splat quenching, liquid quenching, melt quenching or liquisol quenching) is that quenching is carried out from the liquid state at very high cooling rates. However, the shape and size of the quenched product may be different depending on the technique employed. The method of melt spinning- in which a continuous stream of liquid jet is impinged against a conducting wheel rotating at a high speed - has become the most popular since long and continuous ribbons of uniform cross section can be obtained.

Although started only about 20 years ago, the rapid quenching technique has come to be established as a standard method to produce metastable effects. The activity in this area is now world-wide as evidenced by the organization of three international conferences devoted exclusively to this topic. The first one was held in Brela, Yugoslavia (1970), the second in Massachusetts, Cambridge, USA (1975) and the third at Sussex, UK (1978). The fourth is scheduled to be held in Sendai, Japan, during August 1981.

A bibliography on rapid quenching of metallic melts was published in 1973 covering the publications up to 1972. Thus, the present compilation aims at presenting the subsequent literature (1973 onwards) in a chronological order. All the publications of 1980 abailable till the time the book was going to the press are included. The entries presented in this Bibliography are extracted from journals, books and conference proceedings originating from different countries and in several languages. The internal reports are not included unless they are abstracted in standard abstract journals (like Metals Abstracts, Physics Abstracts or Chemical Abstracts). Patents are not included.

Within each year, the entries are arranged alphabetically according to the first-named author. Each reference contains the authors' names (in the sequence they appear in the publication), the full title of the paper, name of the journal, volume, year of publication and page numbers. This is followed, in parentheses, by technique employed (if it is different from rapid quenching from the liquid state) and the properties investigated. The following is the full list of abbreviations used:

Techniques	CD	Chemical Deposition (Electroless)
	ED	Electro Deposition
	II	Ion Implantation
	PS	Plasma Spraying
	Sp	Sputtering
	VD	Vapor Deposition
	At	Atomization
Results	A	Applications
	AB	Annealing Behavior
	AQS	As-Quenched Structure
	AS	Amorphous Solids
	C	Chemical Properties
	E	Electrical Properties
	F	Fabrication
	M	Mechanical Properties
	Mag	Magnetic Properties
	MCS	Metastable Crystalline Intermediate Phases
	PM	Process Mechanisms
	PP	Physical Properties
	QT	Quenching Techniques
	R	Review Articles
	S	Superconductivity
	SMA	Structural Models for Amorphous Phases
	T	Thermal and Thermodynamic Properties
	TSSE	Terminal Solid Solubility Extension

Some of the metastable effects observed in alloys rapidly quenched from the liquid state have also been noticed in alloys produced by other techniques such as vapor deposition, sputtering, etc. Recently, investigations have also begun to compare the effects of different techniques on the constitution and properties of the same alloy. As such, the present bibliography includes the results of other techniques only when that alloy system was also investigated by the melt quenching techniques.

Comprehensive author and alloy indexes are included at the end.

CONTENTS

1973

1. S.C.AGARWAL and H.HERMAN, "Phase Decomposition in Aluminium Alloys Quenched from the Liquid State", in Phase Transitions, eds., H.K.Heinisch, R.Roy and L.E.Cross (Pergamon Press, New York, 1973) pp. 207-22 (R,AB).

2. S.C.AGARWAL and H.HERMAN, "Spinodal Decomposition in Liquid-Quenched Al-22at.%Zn", Scripta Met., 7 (1973) 503-8 (AB).

3. S.C.AGARWAL, M.J.KOCZAK and H.HERMAN, "Precipitation in Liquid-Quenched Al-base Si", Scripta Met., 7 (1973) 365-70 (AB).

4. S.C.AGARWAL, M.J.KOCZAK and H.HERMAN, "Phase Decomposition in Liquid-Quenched Al-28at.%Zn", Scripta Met., 7 (1973) 401-8 (AB).

5. T.R.ANANTHARAMAN and P.RAMACHANDRARAO, "Lattice Imperfections in Liquisol-Quenched Metals and Alloys", The Banaras Metallurgist, 5 (1973) 37-42 (R).

6. M.R.ANSEAU, "Crystallization of $Te_{100-x}Tl_x (15 \leq x \leq 50)$ Obtained by Rapid Cooling from the Liquid State", J. Appl. Phys., 44 (1973) 3357-9 (AB).

7. M.R.ANSEAU, "Application of Ostwald's Rule to Amorphous Materials", Phys. Letters, A43 (1973) 57-9 (VD;E).

8. M.F.ASHBY and J.LOGAN, "The Dislocated Random Network : Dislocation and Disjunctions", Scripta Met., 7 (1973) 513-21 (SMA).

9. E.BABIĆ, E.GIRT, R.KRSNIK, B.LEONTIĆ, M.OČKO, Z.VUČIĆ and I.ZORIĆ, "Microhardness Variation in Al-based 3d Transition Metal Alloys" Phys. Stat. Sol., (a)16 (1973) K21-5 (M).

10. E.BABIĆ, R.KRSNIK and C.RIZZUTO, "The Impurity Resistivity and Resistivity Minima in AlCr and AlMn Alloys", Solid State Commun., 13 (1973) 1027-30 (E).

11. B.G.BAGLEY and F.J.DISALVO, "The Low Temperature Magnetic Properties of Glassy and Crystalline $Pd_{.775}Cu_{.06}Si_{.165}$", in Amorphous Magnetism eds., H.O.Hooper and A.M. de Graaf (Plenum Press, New York, U.S.A. 1973) pp. 143-8 (Mag).

12. B.S.BERRY and W.C.PRITCHET, "Some Physical Properties of Two Amorphous Metallic Alloys", J. Appl. Phys., 44 (1973) 3122-6 (T,M,PP).

13. J.BLÉTRY, "Study of the Splat Cooling of Liquids by Means of a 'Ski-Jump' Piston Apparatus", J. Phys. D: Appl. Phys., 6 (1973) 256-75 (in French) (PM) (See Met. Abstr., 6 (1973) 22-0421).

14. S.K.BOSE and R.KUMAR, "Structure of Rapidly Solidified Aluminium-Silicon Alloys", J. Mater. Sci., 8 (1973) 1795-9 (AQS).

15. Yu.F.BYCHKOV, D.S.KAMENETSKAYA, V.S.KRUGLOV and B.T.SIZOV, "Crystallization of Superconducting Nb-Ga Alloys at High Cooling Rates", Fiz. Met. Metalloved., 36 (1973) 754-9 (in Russian); Phys. Met. Metallogr., 36, No. 4 (1973) 72-7 (TSSE,S).

16. G.S.CARGILL III and R.W.COCHRANE, "Some Structural and Magentic Properties of Amorphous Cobalt-Phosphorus Alloys", in Amorphous Magnetism, eds., H.O.Hooper and A.M. de Graaf (Plenum Press, New York, U.S.A., 1973) pp. 313-20 (ED; AQS,Mag).

17. H.S.CHEN, "Effects of Mixing and Ordering on the Glass Transition in Metallic Glasses of Pd-Ni-P and Pt-Ni-P", in Phase Transitions, eds., H.K.Heinisch, R. Roy and L.E.Cross, (Pergamon Press, New York, 1973) pp. 271-6 (T).

18. H.S.CHEN, "Plastic Flow in Metallic Glasses Under Compression", Scripta Met., 7 (1973) 931-5 (M).

19. H.S.CHEN, "Composition Dependence of the Glass Transition Temperatures of Pd-Ni-P and Pt-Ni-P Glasses", J. Non-Cryst. Solids, 12 (1973) 333-8 (T).

20. H.S.CHEN, "Effects of Composition and Structure on Magnetic Properties of Amorphous Fe-P-C Alloys", Phys. Stat. Sol., (a) 17 (1973) 561-6 (Mag).

21. H.S.CHEN, H.J.LEAMY and M.J.O'BRIEN, "Bending Deformation in Metallic Glasses", Scripta Met., 7 (1973) 415-9 (M).

22. H.S.CHEN and B.K.PARK, "Role of Chemical Bonding in Metallic Glasses", Acta Met., 21 (1973) 395-400 (PP, SMA).

23. H.A.DAVIES, J.AUCOTE and J.B.HULL, "Amorphous Nickel Produced by Splat Quenching", Nature, 246 (1973) 13-4 (AS).

24. H.A.DAVIES and J.B.HULL, "A Non-Crystalline Phase in Splat-Quenched Germanium", Scripta Met., 7 (1973) 637-41 (AS).

25. J.DIXMIER and P.DUWEZ, "Effect of Phosphorus Concentration on the Atomic Distribution Function of Amorphous Pd-Ni-P Alloys", J. Appl. Phys., 44 (1973) 1189-93 (AQS).

26. P.DUHAJ, V.SLÁDEK and P.MRAFKO, "Production of Thin Amorphous Samples Using a Rapidly Rotating Mill Device", Ces. Cas. Fyz. A23 (1973) 617-9 (QT) (in Czech) (See Phys. Abstr., 77 (1974) 17372).

27. M.DUTOIT and H.S.CHEN, "Frequency and Temperature Dependence of Ultrasonic Attenuation in Glassy Pd-Si based Metal Alloys", Appl. Phys. Letters, 23 (1973) 357-8 (PP).

28. P.DUWEZ, "Liquid Quenched Metallic Metastable Alloys", in Proc. Internat. Symp. Fifty Years of Metallurgy, Banaras Hindu University, India (1973) pp. 383-405 (R).

29. P.DUWEZ, "Amorphous Metallic Alloys", Wissen. & Tech. 10 (1973) 311-2 (in German) (R).

30. D.DUŽEVIĆ, A.BONEFAČIĆ and D.KUNSTELJ, "Aluminium-Rich Al-Ni Alloys Rapidly Quenched from the Vapour", Scripta Met., 7 (1973) 883-6 (VD; AQS).

31. R.O.ELLIOTT, A.M.RUSSELL and B.C.GIESSEN, "Retention of a Metastable BCC ε-Pu Solid Solution: ε-Pu(Ti)", J. Mater. Sci., 8 (1973) 1325-30 (TSSE).

32. W.A.ELLIOTT, F.P.GAGLIANO and G.KRAUSS, "Metastable Phases Produced by Laser Melt Quenching", Met. Trans., 4 (1973) 2031-7 (QT, AQS, TSSE).

33. P.FURRER, "Strengthening Mechanisms in Splat-Cooled Aluminium Alloys", in The Microstructure and Design of Alloys, Vol. 1 (The Institute of Metals and The Iron and Steel Institute, London, 1973) pp. 46-50 (M,R).

34. P.FURRER and H.WARLIMONT, "Precipitation of Iron in Rapidly Solidified Aluminium-Iron Alloys", Z. Metallkde., 64 (1973) 236-48 (in German) (AB).

35. P.FURRER and H.WARLIMONT, "Phase Transformation in β-Cu-Al on Extremely Rapid Cooling from the Melt", Z. Metallkde., 64 (1973) 626-34 (in German) (AB).

36. J.J.GILMAN, "Flow via Dislocations in Ideal Glasses", J. Appl. Phys., 44 (1973) 675-9 (M).

37. V.N.GUDZENKO and A.F.POLESYA, "Metastable Phases in the Al-Mg System", Izvest. V. U. Z. Tsvetnaya Met., No. 4 (1973) 144-8 (in Russian)(MCS)(See Met Abstr., 7 (1974) 11-0231).

38. S.P.GUPTA, "Massive Phase Transformation in Splat-Quenched β Ag-Al Alloys", Mater. Sci. & Engg., 12 (1973) 67-77 (AB).

39. R.HASEGAWA, "Anomalies in Amorphous Kondo Alloys", in Amorphous Magnetism, eds., H.O.Hooper and A.M. de Graaf (Plenum Press, New York, 1973) pp. 309-11 (E).

40. R.HASEGAWA and J.A.DERMON, "Electrical Resistivity and Curie Temperature of Amorphous (Fe-Ni)-P-C Alloys", Phys. Letters, A 42 (1973) 407-9 (E,Mag).

41. J.C.HUBERT, F.MOLLARD and B.LUX, "Manufacture of Metallic Wires and Ribbons by the Melt Spin and Melt Drag Processes", Z. Metallkde., 64 (1973) 835-43 (QT).

42. T.ICHIKAWA, "Electron Diffraction Study of the Local Atomic Arrangement in Amorphous Iron and Nickel Films", Phys. Stat. Sol., (a) 19 (1973) 707-16 (VD; AQS).

43. A.K.JENA, B.C.GIESSEN and M.B.BEVER, "On the Metastability of an Au-Sn Phase Prepared by Splat Cooling", Met. Trans., 4 (1973) 279-81 (T).

44. H.JONES, "Splat Cooling and Metastable Phases", Rep. Prog. Phys., 36 (1973) 1425-97 (R).

45. H.JONES and C. SURYANARAYANA, "Rapid Quenching from the Melt: An Annotated Bibliography 1958-72", J. Mater. Sci., 8 (1973) 705-53 (R).

46. H.A.KÄHKÖNEN and M.T.YLI-PENTILA, "Precipitation Phenomena in a Liquisol-Quenched Al-5 at.%Ag-5 at.%Zn Alloy", J. Appl. Cryst., 6 (1973) 412-3 (AB).

47. P.KANTOR, A.REVCOLEVSCHI and R.COLLONGUES, "Preparation of Iron Sesquioxide Glasses by Ultra-Fast Quenching from the Melt (Splat Cooling)", J. Mater. Sci., 8 (1973) 1359-61 (AS).

48. T.Z.KATTAMIS, W.F.BROWER and R.MEHRABIAN, "Microstructure and Segregation in Splat-Cooled Iron-Nickel Alloy", J. Cryst. Growth, 19 (1973) 229-36 (AQS).

49. T.Z.KATTAMIS, Y.V.MURTY and J.A.REFFNER, "Microstructure and Segregation in Splat-Cooled Aluminium-Copper Alloy", J. Cryst. Growth, 19 (1973) 237-41 (AQS).

50. D.KORN, W.MÜRER and G.ZIBOLD, "Temperature Dependence of the Electrical Resistivity in Amorphous Metals", Z. Phys., 260 (1973) 351-60 (VD; E).

51. K.KRANJC and M.STUBIČAR, "X-Ray Diffraction Study of Decomposition in Al-Ag Alloys Quenched from the Melt", Met. Trans., 4 (1973) 2631-8 (AB).

52. R.KRSNIK, E.BABIĆ and C.RIZZUTO, "Phonon Resistivity in Concentrated Aluminium Alloys", Solid State Commun., 12 (1973) 891-5 (E).

53. M.LASOCKA and H.MATYJA, "Conditions of Glassy State Formation in Easily Crystallizing Systems. Discussion Based on the General Shape of the Crystallization Curve", Arch. Hutn. (Poland), 18 (1973) 247-57 (PM)(See Phys. Abstr., 77 (1974) 12873).

54. V.K.C.LIANG and C.C.TSUEI, "Kondo Effect in an Amorphous $Ni_{41}Pd_{41}B_{18}$ Alloy Containing Cr", Phys. Rev., B7 (1973) 3215-25 (E).

55. R.B.LYUKEVICH and A.N.VARICH, "Formation of Phases in Al-Mo Alloys During Non-equilibrium Crystallization", Izvest V.U.Z. Tsvetnaya Met., No. 5 (1973) 131-4 (in Russian) (MCS)(see Met. Abstr., 7(1974) 11-0707).

56. T.B.MASSALSKI, L.F.VASSAMILLET and Y.BIENVENU, "Metastable Phase Relationships in Zinc-rich Copper and Silver Alloys Produced by Rapid Freezing", Acta Met., 21 (1973) 649-61 (TSSE).

57. T.MASUMOTO, "Amorphous Metals", Heat Treatment, 13 (1973) 289-97 (in Japanese) (R).

58. T.MIZOGUCHI, K.YAMAUCHI and H.MIYAJIMA, "Ferromagnetism of Amorphous Iron Alloys", in Amorphous Magnetism, eds., H. O. Hooper and A. M. de Graaf (Plenum Press, New York, U.S.A., 1973) pp. 325-30 (Mag).

59. M.NAGANO, Y.YUKI, T.GOTO and R.IDE, "Melt Spinning of Copper Filament", Sen-i Gakkaishi, 29 (1973) 461-6 (in Japanese)(QT).

60. E.NICOTERA, M.CORCHIA, G.de GEORGI, F.VILLA and M.ANTONINI, "Structural Investigation of Amorphous and Liquid $Ge_{0.175}Te_{0.825}$ by Neutron Scattering", J. Non-Cryst. Solids, 11 (1973) 417-26 (AQS,AS).

61. A.F.POLESYA and V.N.GUDZENKO, "Phase Composition of Ni-Al and Ni-Ga Alloys Obtained by Quenching from the Liquid State", Izvest. V.U.Z., Tsvetnaya Met., No. 2 (1973) 143-8 (in Russian) (MCS)(see Met. Abstr., 6(1973) 11-1000).

62. A.F.POLESYA and V.N.GUDZENKO, "Formation of Metastable States in Cu-Ge and Cu-Si Alloys on Rapidly Cooling from the Liquid State", Izvest. Akad. Nauk SSSR, Metally, No. 5 (1973) 221-4 (in Russian); Russian Met., No. 5 (1973) 152-5 (MCS)(see Met. Abstr., 7(1974) 11-0192).

63. A.F.POLESYA and L.S.SLIPCHENKO, "Formation of Amorphous Phases and Metastable Solid Solutions in Binary Alloys of Titanium and Zirconium with Iron, Nickel and Copper", Izvest. Akad. Nauk. SSSR, Metally, No. 6 (1973) 173-8 (in Russian); Russian Met., No. 6 (1973) 103-7 (AS,TSSE)(see Met. Abstr., 7(1974) 11-0419).

64. D.E.POLK, "Dense Random Packing of Hard Spheres of Two Different Sizes", J. Non-Cryst. Solids, 11 (1973) 381-94 (SMA).

65. D.E.POLK and C.A.PAMPILLO, "The Tensile Strength of an Fe and a Ni-Fe Base Amorphous Alloy", Scripta Met., 7 (1973) 1161-3(M).

66. P.RAMACHANDRARAO, S.LELE and P.RAMA RAO, "On the Impact of Liquisol Quenching on the Structure of a Double HCP Gold-Cadmium-Indium Alloy", J. Mater. Sci., 8 (1973) 451-6 (MCS).

67. P.K.RASTOGI, "Crystallization Behavior of an Amorphous FePC Alloy", J. Mater. Sci., 8 (1973) 140-3 (AB).

68. R.RAY and D.SZYMANSKI, "Electron Diffraction Study of a Non-Crystalline Zr-Ni Phase", Met. Trans., 4 (1973) 1785-90 (AS, AQS).

69. A.V.REVYAKIN et al., "Crystallization of Nickel-Calcium Alloys at High Cooling Rates", Izvest. Akad. Nauk SSSR, Metally, No. 5 (1973) 131-5 (in Russian); Russian Met., No. 5 (1973) 89-91 (AQS)(see Met. Abstr., 7(1974)12-0268).

70. C.RIZZUTO, E.BABIĆ and A.M.STEWART, "Temperature Dependence of the Resistivity in Localized Spin Fluctuation and in Kondo Systems: An Experimental Comparison", J. Phys. F: Metal Phys., 3 (1973) 825-31 (E).

71. R.ROBERGE and H.HERMAN, "Precipitation in Liquid-Quenched Al-Base Ag Alloys", J. Mater. Sci., 8 (1973) 1482-94 (AB).

72. R.ROBERGE and H.HERMAN, "On the Microstructure of Liquid Quenched Monotectoid Al-Zn", J. Inst. Metals, 101 (1973) 339-40 (AQS).

73. A.V.ROMANOVA and V.V.BUKHALENKO, "Crystal Structure of Zinc Quenched from Liuqid State", Fiz. Met. Metalloved., 35 (1973) 1313-4 (in Russian); Phys. Met. Metallogr., 35 No. 6 (1973) 185-6 (AQS).

74. J.F.SADOC, J.DIXMIER and A.GUINIER, "Theoretical Calculation of Dense Random Packing of Equal and Non-equal Sized Hard Spheres - Applications to Amorphous Metallic Alloys", J. Non-Cryst. Solids, 12 (1973) 46-60 (SMA).

75. P.SALVADORI, E.BABIĆ, R.KRSNIK and C.RIZZUTO, "Phonon Resistivity in Zinc Alloys", J. Phys. F: Metal Phys., 3 (1973) L195-8 (E).

76. E.M.SAVITSKY, K.N.IVANOVA, A.V.REVYAKIN and B.D.GLYUZITSKY, "Effect of Ultra Rapid Quenching from the Liquid State on the Structure and Phase Composition of Niobium Alloys", Doklady Akad. Nauk SSSR, 213 (1973) 1289-91 (in Russian); Sov. Phys. Doklady, 18 (1974) 830-1 (AQS)(see Met. Abstr., 7(1974) 11-0491).

77. E.M.SAVITSKY, A.V.REVYAKIN, Yu.V.JEFIMOV, B.D.GLYUZITSKY and V.N.SUMARAKOV, "Extremely Rapid Quenching of Niobium and Vanadium", Doklady Akad. Nauk SSSR, 210 (1973) 405-7 (in Russian) (AQS)(see Met. Abstr., 6(1973) 56-0515).

78. P.H.SHINGU, K.KOBAYASHI, K.SHIMOMURA and R.OZAKI, "Splat Cooling of Aluminium-Silicon Alloys", J. Japan Inst. Metals, 37 (1973) 433-40 (in Japanese) (AQS).

79. C.SURYANARAYANA, "Lattice Parameters of Liquisol-Quenched Aluminium", Phys. Stat. Sol., (a) 18 (1973) K135-7 (AQS).

80. C.SURYANARAYANA, "Constitution of Liquisol-Quenched Al-W Alloys" J. Mater. Sci., 8 (1973) 760-1 (MCS).

81. C.SURYANARAYANA, "Splat Cooling", Sci. Reporter, 10 (1973) 19-21 (R).

82. C.SURYANARAYANA and T.R.ANANTHARAMAN, "On the Structure of a Metastable Phase in the Lead-Bismuth System", Solid State Commun., 12 (1973) 87-8 (MCS).

83. C.SURYANARAYANA and T.R.ANANTHARAMAN, "Metastable Phases in the Aluminium-Germanium System", Z. Metallkde., 64 (1973) 800-4 (TSSE, MCS, AS).

84. T.TAKAMORI and R.ROY, "Rapid Crystallization of SiO_2-Al_2O_3 Glasses", J. Amer. Ceram. Soc., 56 (1973) 639-44 (AB).

85. L.R.TESTARDI, J.T.KRAUSE and H.S.CHEN, "Large Anharmonicity of Amorphous and Crystalline Phases of a Pd-Si Alloy", Phys. Rev., B8 (1973) 4464-9 (M).

86. A.TONEJC, "Annealing and Diffusion Kinetics of Transition Metals Cr, Mn, Fe and Ni in Aluminium", Phil. Mag., 27 (1973) 753-5 (AB).

87. C.C.TSUEI, "Ductile Superconducting Copper-Base Alloys", Science, 180 (1973) 57 (S).

88. C.C.TSUEI, "Magnetic Properties of Amorphous Metallic Alloys", in Amorphous Magnetism, eds. H.O.Hopper and A.M.de Graaf (Plenum Press, New York, U.S.A., 1973) pp. 299-307 (Mag).

89. A.N.VARICH and R.B.LYUKEVICH, "Non-Equilibrium Crystallization of Aluminium-Ruthenium Alloys", Izvest. Akad. Nauk SSSR, Metally, No. 1 (1973) 112-5 (in Russian); Russian Met., No. 1 (1973) 73-5 (AOS)(see Met. Abstr., 6(1973) 12-0932).

90. A.N.VARICH and R.B.LYUKEVICH, "Formation of the Structure of Aluminium - Rhenium Alloys in Non-equilibrium Crystallization", Zhur. Fiz. Khim., 47, No. 4 (1973) 1041 (in Russian) (TSSE) (see Met. Abstr., 6(1973) 12-1230).

91. A.N.VARICH and R.B.LYUKEVICH, "Formation of Intermediate Phases and Thermal Stability of Solid Solutions in the Aluminium - Rhenium System", Zhur. Fiz. Khim., 47, No. 4 (1973) 1042 (in Russian) (TSSE, MCS, AB)(see Met. Abstr., 6(1973) 12-1231).

92. S.YAJIMA, K.OKAMURA and T.SHISHIDO, "Unusual Glass Formation in the Al-Nd-O System", Chem. Letters, No. 7 (1973) 741-2 (AS, AQS).

93. S.YAJIMA, K.OKAMURA and T.SHISHIDO, "Glass Formation in the Ln-Al-O System (Ln: Lanthanoid and Yttrium Elements)", Chem. Letters, No. 12 (1973) 1327-30 (AS, AQS).

94. S.YAJIMA, K.OKAMURA and T.SHISHIDO, "Synthesis of Lanthanoid Aluminates (β-Al_2O_3 Type) Using Arc Plasma Flame", Chem. Letters, No. 12 (1973) 1331-4 (AQS, MCS).

1974

95. S.C.AGARWAL and H.HERMAN, "X-Ray Diffraction Studies of Alloys Quenched from the Liquid State", Siemen's Review, XL1 (1974) 34 (R).

96. H.AHLBORN and D.MERZ, "Aluminium Alloys with Improved Hot Strength and Corrosion Resistance", Aluminium, 50 (1974) 583-7 (M, C).

97. M.R.ANSEAU, "A Comparison of the Limitations of Vapour Quenching and Liquid Quenching in the Presentation of Non-equilibrium Alloys", J. Mater. Sci., 9 (1974) 1189-90 (PM).

98. E.BABIĆ and C.RIZZUTO, "The Linear Variation of the Impurity Resistivity in AlMn, AlCr, ZnFe and Other Dilute Alloys" in Low Temperature Physics-LT13, Vol. 2, eds. K.D.Timmerhaus, W.J. O'Sullivan and E.F.Hammel (Plenum Press, New York, 1974) pp. 484-7 (E).

99. M.BARMATZ and H.S.CHEN, "Young's Modulus and Internal Friction in Metallic Glass Alloys from 1.5 to 300K", Phys. Rev., B9 (1974) 4073-83 (M).

100. A.F.BELYANIN, N.A.BUL'ENKOV, V.P.MARTOVITSKY and T.A.TOPORENSKAYA, "A Study of Magnesium Silicide Alloys Quenched from the Molten State", Tr. Moskov. Inst. Tonkoi Tekhnol., 4 No. 1 (1974) 3-7 (in Russian) (AQS)(see Met. Abstr., 9(1976) 11-0071).

101. S.P.BHAT, T.R.RAMACHANDRAN and A.K.JENA, "Splat Cooling of Aluminium-Manganese Alloys", J. Mater. Sci., 9 (1974) 1759-63 (AB).

102. E.BLANK and E.HORNBOGEN, "The Tempering Behavior of Iron-Copper Alloys after Quenching from Liquid State", Arch. Eisenhütten., 45 (1974) 193-6 (in German) (AB).

103. J.BLÉTRY and J.F.SADOC, "Magnetic Interference Function of Amorphous Co-P Alloys", Phys. Rev. Letters , 33 (1974) 174-5 (ED; Mag).

104. I.N.BOGACHEV, A.G.RUDENKO and L.D.CHUMAKOVA, "Effect of Cooling Conditions on the Substructures of Iron-Manganese Alloys", Izv. Akad. Nauk SSSR, Metally, No. 6 (1974) 167-71 (in Russian) (AQS)(see Met. Abstr., 8(1975) 12-0636).

105. L.M.BUROV and T.I.ANISHCHENKO, "Change in the Width of the X-Ray Diffraction Lines of Rapidly Cooled Al and Al-Cr Alloys Over a Range of Temperatures", Akad. Nauk Ukrain. SSR, Metallofizika, No. 56 (1974) 89-91 (in Russian) (AQS)(see Met. Abstr., 8(1975) 12-1292).

106. Yu.F.BYCHKOV, V.S.KRUGLOV, B.T.SIZOV and E.I.YAKOLEV, "The Effect of Quenching from the Liquid State on the Superconducting Properties and Structure of Nb-Ga and Nb_3(Ge,Al) Alloys", in 'Struktura i Svoistva Sverkhprovodyashch. Materialov', Russian Collection, Nauka, Moscow (1974) pp. 14-19 (in Russian) (S)(see Met. Abstr., 8(1975) 33-1830).

107. G.S.CARGILL III and R.W.COCHRANE, "Amorphous Co-P Alloys: Atomic Arrangements and Magnetic Properties", J. Physique, 35 (1974) C4-269-78 (ED; AQS, Mag).

108. G.S.CARGILL III, R.J.GAMBINO and J.J.CUOMO, "Stripe and Cylindrical Domains in Amorphous Co-P and Co-Ni-P Films", IEEE Trans. Mag., MAG-10 (1974) 803-6 (ED; Mag).

109. M.CHANDRASEKARAN and K.MUKHERJEE, "Substructure of Nickel-Aluminium Martensites in Bulk and Splat-Cooled Specimens", Mater. Sci. & Engg., 14 (1974) 97-9 (AQS).

110. L.A.CHEKANOVA, R.S.ISKHAKOV, G.I.FISH, R.G.KHLEDOPROS and N.S.CHISTYAKOV, "Amorphous State-Polycrystal Phase Transition in Ferromagnetic Co-P Films", Zh. Eskp. & Teor. Fiz. Pis'ma, 20 (1974) 73-6 (in Russian); JETP Letters, 20, No. 2 (1974) 31-2 (CD; AB)(see Phys. Abstr., 78(1975)3835).

111. H.S.CHEN, "The Glass Transition Temperature in Glassy Alloys: Effects of Atomic Sizes and the Heats of Mixing", Acta Met., 22 (1974) 897-900 (T, AS).

112. H.S.CHEN, "Thermodynamic Considerations on the Formation and Stability of Metallic Glasses", Acta Met., 22 (1974) 1505-11 (T, AS).

113. H.S.CHEN and S.Y.CHUANG, "Positron Lifetime Study on the Structure of a Metallic Glass and Interphase Boundaries", Phys. Stat. Sol., (a) 25 (1974) 581-4 (AQS).

114. H.S.CHEN, J.T.KRAUSE and E.A.SIGETY, "Thermal Expansion and Density of Glassy Pd-Ni-P and Pt-Ni-P Alloys", J. Non-Cryst. Solids, 13 (1974) 321-7 (PP).

115. H.S.CHEN and D.E.POLK, "Mechanical Properties of Ni-Fe based Alloy Glasses", J. Non-Cryst. Solids, 15 (1974) 174-8 (M).

116. J.R.CLINTON, E.H.TYLER and H.L.LUO, "Electrical and Magnetic Properties of Au-Ni Alloys", J. Phys. F: Metal Phys., 4 (1974) 1162-9 (E, Mag).

117. R.W.COCHRANE and G.S.CARGILL III, "Magnetization of Amorphous CoP Alloy-Spinwaves in Noncrystalline Ferromagnets", Phys. Rev. Letters, 32 (1974) 476-8 (ED; Mag).

118. J.R.COOPER, Z.VUČIĆ and E.BABIĆ, "The Thermoelectric Power of AlMn Alloys", J. Phys. F: Metal Phys., 4 (1974) 1489-1500 (T).

119. J.COUTURES, F.SIBIEUDE, A.ROUANET, M.FOEX, A.REVCOLEVSCHI and R.COLLONGUES, "Ultra-Fast Quenching of Molten Refractory Oxides. Application to Glass Preparation in Alumina and Gallia-based Systems", Rev. Internat. Hautes Temp. Refract., 11 (1974) 263-7 (QT, AS).

120. H.A.DAVIES and J.B.HULL, "Some Aspects of Splat Quenching in an Inert Atmosphere and of the Formation of Non-Crystalline Phases in Al-17.3 at.% Cu, Germanium and Tellurium", J. Mater. Sci., 9 (1974) 707-17 (AS).

121. H.A.DAVIES, J.AUCOTE and J.B.HULL, "The Kinetics of Formation and Stabilities of Metallic Glasses", Scripta Met., 8 (1974) 1179-89 (PM).

122. J.DIXMIER, "Structure of the Amorphous Metallic Alloys", J. Physique, 35 (1974) C4-11-17 (VD; AQS, SMA).

123. V.I.DOBATKIN and R.M.SIZOVA, "The Non-equilibrium Crystallization of Aluminium-Molybdenum Alloys", Izvest V.U.Z. Tsvetnaya Met., No. 3 (1974) 117-20 (in Russian) (MCS)(see Met. Abstr., 7(1974) 11-0850).

124. P.DUHAJ, V.SLÁDEK and P.MRAFKO, "Electron Microscopic Study of the Crystallization Kinetics of the Amorphous Alloy Pd_{80}-Si_{20}", J. Non-Cryst. Solids, 13 (1974) 341-54 (AB).

125. R.O.ELLIOTT, A.M.RUSSELL and B.C.GIESSEN, "Retention of a Metastable BCC γ-Np Solid Solution: γ-Np (Ga)", Scripta Met., 8 (1974) 1335-40 (TSSE).

126. P.L.FERRAGLIO and K.MUKHERJEE, "The Dynamics of Nucleation and Growth of a Thermoelastic Martensite in a Splat-Quenched Au-47.5 at.%Cd Alloy", Acta Met., 22 (1974) 835-45 (AQS).

127. V.A.FILONENKO, "Structure of Binary Eutectics in the Molten State", Izvest. Akad. Nauk SSSR, Metally, No. 1 (1974) 182-8 (in Russian)(see Met. Abstr., 7(1974) 11-0437).

128. H.FUJIMORI, T.MASUMOTO, Y.OBI and M.KIKUCHI, "On the Magnetization Process in an Fe-P-C Amorphous Ferromagnet", Japan. J. Appl. Phys., 13 (1974) 1889-90 (Mag).

129. P.FURRER, "Extreme Rapid Cooling of Metals", VDI Zeit., 116 (1974) 1297-1301 (R)(see Phys. Abstr., 78(1975) 16159).

130. B.C.GIESSEN, V.O.STRUEBING and R.O.ELLIOTT, "Metastable FCC Solid Solutions in Ce-Pu Alloys Quenched Rapidly from the Melt", in Proc. 11th Rare Earth Res. Conf., Traverse City, Michigan, Vol. 2, eds. J.M.Haschke and H.A.Bick (1974) pp. 632-41 (TSSE).

131. E.GIRT, E.BABIĆ, Z.GATALO, A.KURSUMOVIČ and B.LEONTIĆ, "Modification of Seemann-Bohlin X-Ray Camera for the Study of Metastable States of Ultrarapidly Quenched Samples", J. Phys. E: Sci. Instr., 7 (1974) 354-5 (AQS).

132. E.GIRT, B.LEONTIĆ and A.KURSUMOVIČ, "An Apparatus for the Combined Investigation of Ultrarapidly Quenched Samples", J. Phys. E: Sci. Instrum., 7 (1974) 243-5 (QT).

133. C.V.GOKULARATHNAM, "Structure of Metallic Glasses", J. Mater. Sci., 9 (1974) 673-82 (R).

134. T.GOTO, J.KYUWA, Y.YUKI and M.NAGANO, "The Melt Spinning of the Stainless Steel Fiber Conjugated with Glass", Sen-i Gakkaishi, 30 (1974) 530-6 (in Japanese)(AQS).

135. T.GOTO, Y.YUKI, M.NAGANO, T.ODA and H.TAKAI, "The Melt Spinning of the Low-Carbon Steel Fiber Conjugated with Glass", Sen-i Gakkaishi, 30 (1974) 381-6 (in Japanese) (AQS).

136. M.M.HALL, Jr., "Apparatus for Rapid Quenching in Liquid Metals", Met. Trans., 5 (1974) 955-8 (QT).

137. S.R.HERD and P.CHAUDHARI, "On the Question of Microcrystallites in Some Amorphous Materials. An Electron Microscope Investigation", Phys. Stat. Sol., (a) 26 (1974) 627-42 (VD; AQS).

138. C.E.HIGGS, "The Effect of Heat Treatment on the Structure and Hardness of an Electrolessly Deposited Ni-P Alloy ", Surf. Treat., 2 (1974) 315-26 (CD; AQS, M).

139. T.IKEDA, S.TAGA and S.NISHI, "Effects of Solidification Rate and Thickness on Solidified Structures of Rapidly Solidified Al-3% Mn Alloy", J. Japan Inst. Metals, 38 (1974) 462-71 (in Japanese) (AQS)(see Met. Abstr., 8(1975) 11-0141).

140. M.H.JACOBS, A.G.DOGGETT and M.J.STOWELL, "The Microstructures of Al-8wt.%Fe-based Alloys Prepared by Rapid Quenching from the Liquid State", J. Mater. Sci., 9 (1974) 1631-43 (AQS).

141. W.L.JOHNSON and S.J.POON, "Superconductivity in Metastable Simple Cubic Alloys", J. Appl. Phys., 45 (1974) 3683-4 (S).

142. H.A.KÄHKÖNEN and E.LAINE, "Small-Angle X-ray Study of Splat-Cooled Cadmium", J. Mater. Sci., 9 (1974) 1190-2 (AQS).

143. A.KAHN, J.LIVAGE and R.COLLONGUES, "ESR of V^{4+} in Amorphous and Crystalline V_2O_5", Phys. Stat. Sol., (a) 26 (1974) 175-9 (PP).

144. T.NANEYOSHI, "On the Resistivity Minimum in Amorphous Ferromagnets", Phys. Stat. Sol., (b) 66 (1974) K1-3 (E).

145. M.M.KARNOWSKY, M.MOSS and C.STEIN, "The Activation Energy for the Precipitation of Al_6Mn in Plasma-Sprayed Al-2.9 w/o Mn Alloy", Scripta Met., 8 (1974) 513-8 (PS; AB).

146. N.KAZAMA, T.MASUMOTO and H.WATANABE, "Spin-Wave Excitation in Amorphous Fe-P-C Alloys", J. Phys. Soc. Japan, 37 (1974) 1171 (Mag).

147. A.KIRIN and A.BONEFAČIĆ, "X-Ray Line-Broadening Study of Splat-Cooled Al, Al-Sn and Al-Ni Solid Solutions", J. Phys. F: Metal Phys., 4 (1974) 1608-17 (AQS).

148. M.LASOCKA and H.MATYJA, "Studies of Structural Relaxation of Some Chalcogenide Binary Glasses Using D.S.C. Measurements", J. Non-Cryst. Solids, 14 (1974) 41-7 (T).

149. M.LEBO and N.J.GRANT, "Structure and Properties of a Splat-Cooled 2024 Aluminium Alloy", Met. Trans., 5 (1974) 1547-55 (AQS, M).

150. J.LOGAN, "Defects in Glassy Pd-Si Alloys", Scripta Met., 8 (1974) 727-32 (AQS).

151. J.LOGAN and M.F.ASHBY, "Mechanical Properties of Two Metallic Glasses", Acta Met., 22 (1974) 1047-54 (ED; M).

152. A.MADHUKAR, "Indirect Exchange Mechanism of Magnetic Ordering in Amorphous Alloys", J. Physique, 35 (1974) C4-295-8 (Mag).

153. A.MADHUKAR and R.HASEGAWA, "Resistivity Minimum in Amorphous Ferromagnets", J. Physique, 35 (1974) C4-291-3 (E).

154. A.MADHUKAR and R.HASEGAWA, "Mechanism for Resistivity Minimum in Amorphous Ferromagnets", Solid State Commun., 14 (1974) 61-4 (E).

155. G.MANFRE, G.SERVI and C.RUFFINO, "Copper Microwires Spun from the Melt", J. Mater. Sci., 9 (1974) 74-80 (QT).

156. R.E.MARINGER and C.E.MOBLEY, "Casting of Metallic Filament and Fiber", J. Vac. Sci. & Tech., 11 (1974) 1067-71 (QT).

157. T.MASUMOTO, "Strength and Fracture of Amorphous Metals", in Proc. 19th Japan Symp. on Strength, Fracture and Fatigue, 19 (1974) 14 (in Japanese) (R, M).

158. R.MISHIMA, Y.ISHIDA and M.KATO, "Structures and Mössbauer Spectra of Splat-Cooled Al-Fe Alloys", J. Japan Inst. Light Metals, 24 (1974) 204-10 (in Japanese) (AQS)(see Met. Abstr., 8(1975) 14-0233).

159. R.MISHIMA, Y.ISHIDA and M.KATO, "Ageing Behavior and Mössbauer Spectra of Splat-Cooled Al-Fe Alloys", J. Japan Inst. Light Metals, 24 (1974) 211-5 (in Japanese) (AB)(see Met. Abstr., 8 (1975) 14-0234).

160. T.MIZOGUCHI and Y.YAMAUCHI, "Critical Behavior of Amorphous Ferromagnet", J. Physique, 35 (1974) C4-287-90 (Mag).

161. M.MORINAGA, M.MURAKAMI, S.NASU, Y.MURAKAMI, P.H.SHINGU and T.TAOKA, "Supersaturated Solid Solution of Al-5.7 at.%Ag Alloy Prepared by Splat Quenching", J. Mater. Sci., 9 (1974) 1385-7 (TSSE).

162. P.MRAFKO and P.DUHAJ, "X-Ray Study of an Amorphous $Pd_{80}Si_{20}$ Alloy", Phys. Stat. Sol., (a) 22 (1974) 151-5 (AQS).

163. P.MRAFKO and P.DUHAJ, "Analysis of an Aggregate of Hard Spheres", Phys. Stat. Sol., (a) 23 (1974) 583-9 (SMA).

164. M.NAKA, K.HASHIMOTO and T.MASUMOTO, "Corrosion Resistivity of Amorphous Iron Alloys Containing Chromium", J. Japan Inst. Metals, 38 (1974) 835-41 (in Japanese) (C)(see Met. Abstr., 8(1975) 35-1111).

165. S.NASU, U.GONSER, P.H.SHINGU and Y.MURAKAMI, "^{57}Fe Mössbauer Spectra in Splat-Quenched Al-0.5,1.3 and 5 at.%Fe Alloys", J. Phys. F: Metal Phys., 4 (1974) L24-8 (AQS).

166. C.A.PAMPILLO and H.S.CHEN, "Comprehensive Plastic Deformation of a Bulk Metallic Glass", Mater. Sci. & Engg., 13 (1974) 181-8 (M).

167. C.A.PAMPILLO and D.E.POLK, "The Strength and Fracture Characteristics of Fe, Ni-Fe and Ni-base Glasses at Various Temperatures", Acta Met., 22 (1974) 741-9 (M).

168. C.A.PAMPILLO and A.C.REIMSCHUESSEL, "The Fracture Topography of Metallic Glasses", J. Mater. Sci., 9 (1974) 718-24 (M).

169. D.PAN and D.TURNBULL, "Magnetic Properties of Amorphous Co-P Alloys", J. Appl. Phys., 45 (1974) 1406-12 (ED; Mag).

170. D.PAN and D.TURNBULL, "Magnetic Properties of Amorphous Ni-P Alloys", AIP Conf. Proc., 18 (1974) 646-50 (ED; Mag).

171. D.I.PAUL, J.MARTI and L.VALADEZ, "Ferromagnetic Domain Walls in Amorphous Iron Alloys", AIP Conf. Proc., 18 (1974) 1377-81 (Mag).

172. A.F.POLESYA and V.N.GUDZENKO, "Metastable and Amorphous Phases in the System Mn-Si", Izv. Akad. Nauk SSSR, Neorg. Materialy, 10 (1974) 1011-5 (in Russian); Inorganic Mater., 10 (1974) 869-72 (TSSE, MCS, AS)(see Met. Abstr., 7(1974) 11-0760).

173. A.F.POLESYA and L.S.SLIPCHENKO, "Structure of Rapidly Solidified Alloys of the Nickel-Germanium System", Izv. Akad. Nauk. SSSR. Metally, No. 2 (1974) 239 (in Russian) (AQS)(see Met. Abstr., 7(1974) 11-0597).

174. A.F.POLESYA and L.S.SLIPCHENKO, "Structure of Ni-Sb and Fe-Sb Alloys Quenched from the Molten State", Izvest V.U.Z. Chernaya Met., No. 3 (1974) 149-53 (in Russian) (AQS)(see Met. Abstr., 7(1974) 11-0705).

175. A.F.POLESYA and L.S.SLIPCHENKO, "Formation of Metastable Iron-Tin Solid Solutions", Akad. Nauk Ukrain SSR, Metallofizika, No. 51 (1974) 112-5 (in Russian) (TSSE)(see Met. Abstr., 8 (1975) 11-0110).

176. A.F.POLESYA and L.S.SLIPCHENKO, "Structure of Splat-Quenched Iron-Germanium Alloys", Akad. Nauk Ukrain SSR, Metallofizika, No. 50 (1974) 95-8 (in Russian) (AB)(see Met. Abst., 8(1975) 11-0507).

177. D.E.POLK and H.S.CHEN, "Formation and Thermal Propertis of Glassy Ni-Fe based Alloys", J. Non-Cryst. Solids, 15 (1974) 165-73 (AS, T).

178. P.RAMACHANDRARAO and M.LARIDJANI, "A Metastable Phase Al_3Cu_2", J. Mater. Sci., 9 (1974) 434-7 (MCS).

179. R.ROBERGE and H.HERMAN, "Phase Decomposition in Liquid Quenched Al-14 at.%Ag: Small Angle X-Ray Scattering Study", J. Mater. Sci., 9 (1974) 1123-7 (AB).

180. I.V.SALLI, V.Z.DOLINSKAYA and L.V.SAMOYLENKO, "Influence of Cooling Conditions on Structure Formation in Lead-Antimony Alloys", Fiz. Met. Metalloved., 38 (1974) 132-8 (in Russian); Phys. Met. Metallogr., 38, No. 1 (1974) 118-23 (AQS).

181. I.V.SALLI, V.F.KATKOV and A.K.KUSHNEREVA, "The Effect of Non-equilibrium Crystallization on the Yield Point of Al-Mn and Al-Zr Alloys", Fiz. Khim. Obrab. Mat., No. 1 (1974) 151-3 (in Russian) (M)(see Met. Abstr., 7(1974) 31-3025).

182. I.V.SALLI and L.V.SAMOYLENKO, "On the Growth Forms of Tin Antimonide Crystals in Alloys of the Peritectic Type", Akad. Nauk Ukrain, SSR, Metallofizika, No. 56 (1974) 68-71 (AQS) (see Met. Abstr., 8(1975) 12-1290).

183. P.R.SARODE and A.R.CHETAL, "Metastable Solid Solutions in Lead-Tin Alloys", Current Sci. (India), 43 (1974) 339 (TSSE).

184. E.M.SAVITSKY, K.N.IVANOVA, A.V.REVYAKIN, B.D.GLYUZITSKY and P.A.KHANDAROV, "Effect of Extremely Fast Quenching and Ageing on the Structure and Properties of Nb-W-Zr (Hf)-C Alloys", Fiz. Met. Metalloved., 38 (1974) 117-25 (in Russian); Phys. Met. Metallogr., 38, No. 1 (1974) 103-11 (AB, AQS, M).

185. E.M.SAVITSKY, A.V.REVYAKIN, Yu.V.JEFIMOV, B.D.GLYUZITSKY and V.N.SUMARAKOV, "The Effect of the Rate of Quenching and Nitrogen Content on the Structure and Superconducting Properties of Vanadium and Niobium and Some of Their Alloys", in Struktura i Svoistva Sverkhprovodyashch. Materiolov, Russian Collection, Nauka, Moscow (1974) pp. 145-56 (AQS, S) (see Met. Abstr., 8(1975) 33-1843).

186. T.SAWAKI, T.TAKEUCHI and S.TAKASU, "Relation Between the Crystallization and Strength of Duralumin Filament Produced by Melt Spinning", J.T.M.S., 27, No. 3 (1974) 72-5 (in Japanese) (AB, M).

187. M.G.SCOTT, "The Effect of a Glass Substrate on Cooling Rate of Splat Quenching", J. Mater. Sci., 9 (1974) 1372-4 (PM).

188. M.G.SCOTT, "Comments on 'Metastable Phases in the Aluminium-Germanium System'", Z. Metallkde., 65 (1974) 563-5 (TSSE, MCS, AS)(see entry 83).

189. P.H.SHINGU, K.KOBAYASHI, K.SHIMOMURA and R.OZAKI, "An Amorphous Phase in a Splat-Cooled Fe-3.8 wt.% C Alloy", Scripta Met., 8 (1974) 1317-9 (AS).

190. F.SPAEPEN and D.TURNBULL, "A Mechanism for the Flow and Fractur of Metallic Glasses", Scripta Met., 8 (1974) 563-8 (M).

191. C.SURYANARAYANA, "A New Metastable Phase in the Silver-Silicon System", J.Less-Common Metals, 35 (1974) 347-52 (MCS).

192. C.SURYANARAYANA and T.R.ANANTHARAMAN, "On the Crystal Structure of a Non-equilibrium Phase in the Gold-Silicon System", Mater. Sci. & Engg., 13 (1974) 73-81 (MCS).

193. T.SUZUKI and A.M.ANTHONY, "Rapid Quenching on the Binary Systems of High Temperature Oxides", Mater. Res. Bull., 9 (1974) 745-53 (AS, AQS).

194. F.R.SZOFRAN, J.M.WEYMOUTH, G.R.GRUZALSKI, D.J.SELLMYER, R.RAY and B.C.GIESSEN, "Spin Fluctuation and Spin-spin Interactions in Amorphous Metallic Alloys", AIP Conf. Proc., 18 (1974) 282-6 (Mag).

195. K.TAMURA, J.FUKUSHIMA, H.ENDO K.KISHI, S.IKEDA and S.MINOMURA, "X-Ray Photoemission Spectra of Valence Bands in Amorphous Ge-Ni, Ge-Fe and Ge-Au Alloys", J. Phys. Soc. Japan, 36 (1974) 565-71 (VD; PP).

196. Yu.N.TARAN, I.S.MIROSHNICHENKO and I.M.GALUSHKO, "Effect of Cooling Rate on the Structure of Eutectic Alloys", Akad. Nauk Ukrain. SSR, Metallofizika, No. 56 (1974) 77-83 (in Russian) (AQS)(see Met. Abstr., 8(1975) 12-1291).

197. G.THURSFIELD and M.J.STOWELL, "Mechanical Properties of Al-8wt.%Fe-based Alloys Prepared by Rapid Quenching from the Liquid State", J. Mater. Sci., 9 (1974) 1644-60 (M).

198. A.M.TONEJC, A.TONEJC and A.BONEFAČIĆ, "Nonequilibrium Phases in Al-rich Al-Pt Alloys", J. Mater. Sci., 9 (1974) 523-6 (TSSE, MCS).

199. C.C.TSUEI and W.L.JOHNSON, "Superconductivity in Metal-Semiconductor Eutectic Alloys", Phys. Rev., B9 (1974) 4742-51 (S).

200. D.TURNBULL, "Amorphous Solid Formation and Interstitial Solution Behavior in Metallic Alloy Systems", J. Physique, 35 (1974) C4-1-10 (AS).

201. Y.S.TYAN and L.E.TOTH, "Microstructure of Amorphous and Crystalline Electrodeposited Ni-P and Co-P Alloys", J. Electron. Mater., 3 (1974) 791-820 (ED; AQS).

202. N.I.VARICH, A.B.LYSENKO, V.I.TKACH and A.A.YAKUNIN, "Formation and Decomposition of the 'X-Ray Amorphous' Phase in Pr-Ag Alloys", Zhur. Fiz. Khim., 48 (1974) 1706-08 (in Russian) (AS, AB)(see Met. Abstr., 8(1975) 11-0134).

203. V.M.VOZDVIZHENSKII and S.S.VOEVODINA, "The Formation of Solid Solutions in Al-Alloys with High Rates of Cooling", Sb. Tr. Rybinsk Aviats, Teknol.Inst., No. 2 (1974) 148-61 (TSSE)(see Met. Abstr., 9(1976) 11-0199).

204. J.C.A.VREESWIJK, R.G.GOSSINK and J.M.STEVELS, "Nucleation Kinetics and Critical Cooling Rate of Glass-Forming Liquids", J. Non-Cryst. Solids, 16 (1974) 15-27 (PM).

205. R.WANG and Y.B.KIM, "Metastable Solubility Limits of Primary HCP Solid Solutions Quenched from the Liquid State", Met. Trans., 5 (1974) 1973-7 (TSSE).

206. R.M.WHITE, "Random Network Model for Amorphous Alloys", J. Non-Cryst. Solids, 16 (1974) 387-98 (SMA).

207. H.WIESNER and J.SCHNEIDER, "Magnetic Properties of Amorphous Fe-P Alloys Containing Ga, Ge and As", Phys. Stat. Sol., (a) 26 (1974) 71-5 (Mag).

208. C.A.WILLIAMS and H.JONES, "Simple Device for Quenching from the Melt", Metals Tech., 1 (1974) 202-3 (QT).

209. J.V.WOOD and R.W.K.HONEYCOMBE, "The Structure of Splat-Cooled Fe-20% Cr-25% Ni Austenitic Steel", J. Mater. Sci., 9 (1974) 1183-8 (AQS).

210. S.YAJIMA, K.OKAMURA and T.SHISHIDO, "Crystallization of an Amorphous Oxide in La-Nb-O System", Chem. Letters, No. 3 (1974) 221-2 (AB).

211. S.YAJIMA, K.OKAMURA and T.SHISHIDO, "Crystallization of $3Ln_2O_3$-$5Ga_2O_3$ Glasses", Chem. Letters, No. 6 (1974) 545-8 (AB).

212. S.YAJIMA, K.OKAMURA and T.SHISHIDO, "$3Gd_2O_3$-$5Fe_2O_3$ Glass Obtained by Rapid Quenching Apparatus Using Laser Beam", Chem. Letters, No. 12 (1974) 1531-4 (AS).

213. S.YAJIMA, K.OKAMURA and T.SHISHIDO, "Formation of Unusual Glasses based on Ln_2O_3", in Proc. 11th Rare Earth Research Conf., Traverse City, Michigan, eds. J.M.Haschke and H.A.Bick (1974), pp 568-77 (AS).

214. M.YOSHIDA and S.NAKAMURA, "The Amorphous Phases Obtained by Ultra-Rapid Solidification", J. Soc. Mater. Sci. Japan, 23 (1974) 605-14 (in Japanese) (AS)(see Met. Abstr., 8(1975) 12-0766).

215. T.YOSHIIE, K.YAMAKAWA and F.E.FUJITA, "Electrical Resistivity Minimum in Amorphous Iron Films", J. Phys. Soc. Japan, 37 (1974) 572 (VD; E).

216. A.ZENTKO, P.DUHAJ, P.MARKO, L.POTOCKÝ and T.TIMA, "Influence of Concentration of Cobalt on the Magnetic Properties of Amorphous $Pd_{80-x}Co_xSi_{20}$ Alloys", Phys. Stat. Sol., (a) 24 (1974) K99-101 (Mag).

217. A.ZENTKO, P.DUHAJ, L.POTOCKÝ, T.TIMA and P.MARKO, "The Low Temperature Minimum of the Electrical Resistivity of Amorphous $Pd_{80-x}Co_xSi_{20}$ Alloys", Phys. Stat. Sol., (a) 24 (1974) K103-5 (E).

1975

218. N.E.ALEKSEEVSKY and E.P.KRASNOPEROV, "Effect of Quenching on the Critical Temperatures of the Niobium-Germanide-Niobium Aluminide (Superconducting) System", Fiz. Met. Metalloved, 39 (1975) 872-3 (in Russian); Phys. Met. Metallogr., 39, No. 4 (1975) 176-8 (S).

219. M.R.ANSEAU and L.DANGUY, "Comparison of the Structural Characteristics of Non-equilibrium Alloys Prepared by Splat Cooling or Vapour Quenching", Acta Technica Belgica-Metallurgie, 15 (1975) 166-8 (in French) (PM)(see Met. Abstr., 9(1976) 11-0546).

220. J.D.AXE, L.PASSELL and C.C.TSUEI, "Spin Waves in Amorphous Metallic Ferromagnet", AIP Conf. Proc., 24 (1975) 119-20 (Mag).

221. E.BABIĆ, R.KRSNIK and M.OČKO, "Low Temperature Resistivity of Al V Alloys", in Low Temperature Physics-LT14, Vol. 3, eds. M.Krusius and M.Vuorio (North-Holland Elsevier, Amsterdam, 1975) pp. 102-5 (E).

222. E.BABIĆ, R.KRSNIK and C.RIZZUTO, "The Impurity Resistivity of ZnFe Alloys", J. Low. Temp. Phys., 21 (1975) 243-55 (E).

223. B.G.BAGLEY and E.M.VOGEL, "The Crystallization Kinetics of Glassy $Pd_{0.775}Cu_{0.06}Si_{0.165}$", J. Non-Cryst. Solids, 18 (1975) 29-32 (AB).

224. J.J.BECKER, "Magnetization Reversal Behaviour in an Amorphous Alloy", IEEE Trans. Mag., MAG-11 (1975) 1326-8 (Mag).

225. B.S.BERRY and W.C.PRITCHET, "Magnetic Annealing and Directional Ordering of an Amorphous Ferromagnetic Alloy", Phys. Rev. Letters, 34 (1975) 1022-5 (Mag).

226. J.BLÉTRY and J.F.SADOC, "Determination of the Three Partial Interference Functions of an Amorphous Co-P Ferromagnet by Polarized Neutron Scattering", J. Phys. F: Metal Phys., 5 (1975) L110-7 (ÉD; AQS).

227. A.BONEFAČIĆ, M.KERENOVIĆ, A.KIRIN and D.KUNSTELJ, "Segregation of Solutes in Al-Ni and Al-Sn and its Influence on their Mechanical Properties", J. Mater. Sci., 10 (1975) 243-51 (AQS, M).

228. R.W.CAHN, "Splat Quenching", Metal Sci., 9 (1975) 518-9 (R).

229. G.S.CARGILL III, "Structure of Metallic Alloy Glasses", in Solid State Phys, Vol. 30 eds. H.Ehrenreich, F.Seitz and D.Turnbull (Academic Press, New York, 1975) pp. 227-320 (R).

230. G.S.CARGILL III, "Ferromagnetism in Amorphous Solids", AIP Conf. Proc., 24 (1975) 138-44 (ED; Mag).

231. G.S.CARGILL III, "Atomic Scale Structure and Magnetic Properties of Some Metallic Alloy Glasses", J. Physique, 36 (1975) C2-71-2 (ED; AQS, Mag).

232. H.S.CHEN, "Effects of Cold Rolling on Young's Modulus and Structure of Metallic Glasses", Scripta Met., 9 (1975) 411-5 (PM, M).

233. H.S.CHEN and S.Y.CHUANG, "Structural Investigations of Cold-Rolled Metallic Glasses Using Positron Annihilation Methods", Appl. Phys. Letters, 27 (1975) 316-7 (AQS).

234. H.S.CHEN and S.Y.CHUANG, "Structure Study of a Metallic Glass Using Positron Annihilation Methods", J. Electron Mater., 4 (1975) 783-9 (AQS).

235. H.S.CHEN, S.D.FERRIS, E.M.GYORGY, H.J.LEAMY and R.C.SHERWOOD, "Field Heat Treatment of Ferromagnetic Metallic Glasses", Appl. Phys. Letters, 26 (1975) 405-6 (Mag).

236. H.S.CHEN, J.T.KRAUSE and E.COLEMAN, "Elastic Constants, Hardness and their Implications to Flow Properties of Metallic Glasses", J. Non-Cryst. Solids, 18 (1975) 157-71 (M).

237. H.S.CHEN, J.T.KRAUSE and E.COLEMAN, "Young's Modulus and Hardness of Metallic $(Pd_{1-x}Ni_x)_{0.835}Si_{0.165}$ Glasses", Scripta Met., 9 (1975) 787-9 (M).

238. C.L.CHIEN, J.C.WALKER and R.HASEGAWA, "Observation of RF-Induced Sideband Effects in an Amorphous Magnetic Material", AIP Conf. Proc., 24 (1975) 127-8 (Mag).

239. C.P.CHOU and F.SPAEPEN, "Some Mechanical Properties of Phase Separated $Pd_{0.74}Au_{0.08}Si_{0.18}$ Metallic Glasses", Acta Met., 23 (1975) 609-13 (M).

240. C.P.CHOU and D.TURNBULL, "Transformation Behavior of Pd-Au-Si Metallic Glasses", J. Non-Cryst. Solids, 17 (1975) 169-88 (AB).

241. S.Y.CHUANG, S.J.TAO and H.S.CHEN, "Investigation of the Electronic Structure of a Glassy $Pd_{.775}Cu_{.06}Si_{.165}$ Alloy by Angular Correlation Studies of Positron Annihilation Radiation", J. Phys. F: Metal Phys., 5 (1975) 1681-6 (AQS).

242. M.CHYCZEWSKI, "Investigation of the Possibility of Formation of Metastable Intermediate Phases in Alloys of Antimony, Bismuth and Tellurium", Z. Metallkde., 66 (1975) 119-20 (MCS).

243. R.W.COCHRANE, R.HARRIS, M.PLISCHKE, D.ZOBIN and M.J.ZUCKERMANN, "Mössbauer Absorption Spectra in Amorphous Metallic Alloys", Phys. Rev., B12 (1975) 1969-70 (AQS).

244. R.W.COCHRANE, R.HARRIS, M.PLISCHKE, D.ZOBIN and M.J.ZUCKERMANN, "Magnetic Properties of a Random Close Packed Amorphous Alloy", J. Phys. F: Metal Phys., 5 (1975) 763-73 (Mag).

245. R.W.COCHRANE, R.HARRIS, J.O.STROM-OLSEN and M.J.ZUCKERMANN, "Structural Manifestations in Amorphous Alloys: Resistance Minima", Phys. Rev. Letters, 35 (1975) 676-9 (ED; E).

246. R.W.COCHRANE, R.HARRIS, J.O.STROM-OLSEN and M.J.ZUCKERMANN, "Resistance Anomalies in Amorphous Metals", in Low Temperature Physics-LT14, Vol. 3, eds. M.Krusius and M.Vuorio (North-Holland Elsevier, Amsterdam, 1975) pp. 118-21 (E)(see Chem. Abstr., 85(1975) 71153t).

247. O.COLLET, A.FONTAINE and A.GUINIER, "Structure of Al-Fe Alloys Obtained by Coevaporation", Mater. Sci. & Engg., 19 (1975) 277-81 (in French) (VD; AQS).

248. L.DAGENS, "Origin of the Contrasts around the Small Ordered Θ" Precipitates in Al-3 at.%Fe (Splat-Cooled Alloy)", J. Physique, 36 (1975) 531-44 (in French) (AB)(see Met. Abstr., 9(1976) 14-0203).

249. E.DARTYGE, "Formation of Amorphous Material", in 'Radiation Damage Processes in Materials', Noordhoff Internat. Pub., Groningen, Netherlands, 1975, pp. 389-404 (R)(see Phys. Abstr. 79(1976) 9112).

250. H.A.DAVIES, "The Kinetics of Formation of a Au-Ge-Si Metallic Glass", J. Non-Cryst. Solids, 17 (1975) 266-72 (PM).

251. H.A.DAVIES "Glassy Alloys Ready to Take Off", New Scientist, 65 (1975) 791 (R).

252. H.A.DAVIES and B.G.LEWIS, "A Generalised Kinetic Approach to Metallic Glass Formation", Scripta Met., 9 (1975) 1107-12 (PM).

253. L.A.DAVIS, "Plastic Instability in a Metallic Glass", Scripta Met., 9 (1975) 339-41 (M).

254. L.A.DAVIS, "Hardness/Strength Ratio of Metallic Glasses", Scripta Met., 9 (1975) 431-5 (M).

255. L.A.DAVIS, "Fracture of Ni-Fe base Metallic Glasses", J. Mater. Sci., 10 (1975) 1557-64 (M).

256. L.A.DAVIS and S.KAVESH, "Deformation and Fracture of an Amorphous Metallic Alloy at High Pressure", J. Mater. Sci., 10 (1975) 453-9 (M).

257. J.DIXMIER, J.BLÉTRY and J.F.SADOC, "Neutron Diffraction Study of the Structure of Amorphous Metallic Alloys", J. Physique, 36 (1975) C2-65-8 (in French) (VD; AQS).

258. M.DOYAMA, S.TANIGAWA, K.KURIBAYASHI, H.FUKUSHIMA, K.HINODE and F.SAITO,"Positron Annihilation in Glassy Metals", J. Phys. F: Metal Phys., 5 (1975) L230-2 (ED; AQS).

259. A.M.DURACHENKO, A.I.DUKHIN, V.T.BORISOV and I.N.GOLIKOV, "Conditions Governing the Formation of Amorphous Alloys on Quenching from the Liquid Phase and their Subsequent Crystallization", Doklady Akad. Nauk SSSR, 225 (1975) 293-5 (in Russian); Sov. Phys. Doklady, 20 (1975) 772-4 (AS, AB)(see Met. Abstr., 9(1976) 12-0563).

260. P.DUWEZ, "Liquid Quenched Metallic Metastable Alloys", Acta Technica Belgica-Metallurgie, 15 (1975) 157-65 (R)(see Met. Abstr., 9(1976) 11-0545).

261. D.DUŽEVIĆ, "Vapour-Quenched Al-Ni Alloy Thin Film in Concentration range 12.5 to 75.4 at.%Ni", Scripta Met., 9 (1975) 543-6 (VD; AQS).

262. T.EGAMI, P.J.FLANDERS and C.D.GRAHAM, Jr., "Amorphous Alloys as Soft Magnetic Materials", AIP Conf. Proc., 24 (1975) 697-701 (Mag).

263. T.EGAMI, P.J.FLANDERS and C.D.GRAHAM, Jr., "Low-Field Magnetic Properties of Ferromagnetic Amorphous Alloys", Appl. Phys. Letters, 26 (1975) 128-30 (Mag).

264. W.C.EMMENS, J.VRIJEN and S.RADELAAR, "Crystallization of Amorphous $Pd_{0.75}Ag_{0.05}Si_{0.20}$ Under Hydrostatic Stress", J. Non-Cryst. Solids, 18 (1975) 299-302 (AB).

265. H.FAHLENBRACH, "Amorphous Magnetic Materials and Their Magnetic Properties", Elektron Ind., 6 (1975) 217-20 (in German) (R, Mag)(see Phys. Abstr., 79(1976) 36104).

266. J.L.FINNEY, "The Structures of Laboratory and Computer-Built Random Packing", J. Physique, 36 (1975) C2-1-11 (SMA).

267. P.J.FLANDERS, C.D.GRAHAM, Jr. and T.EGAMI, "Magnetic Properties of Amorphous Magnetic Alloys", IEEE Trans. Mag., MAG-11 (1975) 1323-5 (Mag).

268. M.FOEX, F.SIBIUDE, A.ROUANET and D.HERNANDEZ, "Crystal-Chemical Effect of Splat Cooling on a 30 mol % CeO_2 70 mol % La_2O_3 Mixed Oxide", J. Mater. Sci., 10 (1975) 1255-7 (AQS).

269. A.FONTAINE and A.GUINIER, "Study of Splat-Cooled Aluminium-Iron Alloys", Phil. Mag., 31 (1975) 65-85 (in French) (AQS).

270. A.FONTAINE and A.GUINIER, "Precipitation in Splat-Cooled Al-3 at.%Fe Alloy", Phil. Mag., 31 (1975) 839-53 (in French) (AB).

271. A.I.GAL, V.V.GAL and R.A.FANIN, "Disintegration of the Metastable Gamma-Phase in Fe-Al Powder Compounds Prepared at High Cooling Rates", Phys. Stat. Sol., (a) 32 (1975) K115-8 (AB).

272. B.C.GIESSEN, V.O.STRUEBING and R.O.ELLIOTT, "Metastable F.C.C. Pu-Ce Solid Solutions", Mater. Sci. & Engg., 18 (1975) 239-43 (TSSE).

273. J.J.GILMAN, "Metallic Glasses", Phys. Today, 28 No. 5 (1975) 46-53 (R).

274. J.J.GILMAN, "Mechanical Behaviour of Metallic Glasses", J. Appl. Phys., 46 (1975) 1625-33 (R).

275. V.M.GLAZOV, R.A.AKOPYAN, A.V.YEVDOKIMOV and G.Yu.FILIPPOVA, "The Diffusionless Crystallization of Alloys", Zavod. Lab., 41 (1975) 560-3 (in Russian) (QT)(see Met. Abstr., 8(1975) 21-0565).

276. C.H.L.GOODMAN, "Strained Mixed-Cluster Model for Glass Structure", Nature, 257 (1975) 370-2 (SMA).

277. T.GOTO, M.NAGANO and K.TANAKA, "The Melt Spinning of Iron Filament", Sen-i Gakkaishi, 31 (1975) 121-7 (in Japanese) (AQS).

278. T.GOTO, M.NAGANO, Y.YUKI, T.ODA and H.TAKAI, "Exothermic Behavior on Heating of Copper Fiber Manufactured by the Method of Glass Conjugated Melt Spinning", Sen-i Gakkaishi, 31 (1975) 104-8 (in Japanese) (AQS).

279. W.GRUHL, B.GRZEMBA and W.HILLER, "Experiments on Electron Beam Melting and Enrichment of Aluminum Alloys", in Proc. 6 Internat. Leichtmetaltagung, Leoben/Wien(Aluminium Verlag, Dusseldorf, 1975)pp.33-5 (in German) (TSSE)(see Met. Abstr., 9(1976) 56-0451).

280. P.J.GRUNDY, S.S.NANDRA and A.ALI, "The Magnetization of Amorphous Films of Some Ferromagnetic Transition Metals and Their Alloys", IEEE Trans. Mag., MAG-11 (1975) 1329-31 (VD; Mag).

281. V.N.GUDZENKO and A.F.POLESYA, "Metastable Phases in Mn-Base Alloys", Izv. Akad. Nauk. SSSR, Met., No. 5 (1975) 192-5 (in Russian) (MCS)(see Met. Abstr., 9(1976) 11-0250).

282. V.N.GUDZENKO and A.F.POLESYA, "Structure of Zr-Al Alloys Rapidly Cooled from the Liquid State", Fiz. Met. Metalloved., 39 (1975) 1313-5 (in Russian); Phys. Met. Metallogr., 39 No. 6 (1975) 177-9 (AQS).

283. D.GUPTA, K.N.TU and K.W.ASAI, "Diffusion in the Amorphous Phase of Pd-19 at.%Si Metallic Alloy", Phys. Rev. Letters, 35 (1975) 796-9 (PP).

284. R.HASEGAWA and R.C.TAYLOR, "Magnetization of Amorphous Gd-Co-Ni Films", J. Appl. Phys., 46 (1975) 3606-8 (Mag).

285. A.HEIDEMANN, "Hyperfine Interaction in Amorphous Ferromagnetic Co-P Alloys Measured by Inelastic Scattering of Neutrons", Z. Physik, B20 (1975) 385-9 (Mag).

286. L.V.HEIMENDAHL, "Metallic Glasses as Relaxed Bernal Structures", J. Phys. F: Metal Phys., 5 (1975) L141-5 (SMA).

287. L.V.HEIMENDAHL and M.F.THORPE, "Phonons in Metallic Glasses", J. Phys. F: Metal Phys., 5 (1975) L87-91 (AS).

288. T.ICHIKAWA, "The Assembly of Hard Spheres as a Structure Model of Amorphous Iron", Phys. Stat. Sol., (a) 29 (1975) 293-302 (SMA).

289. W.L.JOHNSON and S.J.POON, "Superconductivity of New Metastable Phases of Nb-Ir, Ta-Ir and Ta-Rh", J. Less-Common Metals, 42 (1975) 355-8 (S).

290. W.J.JOHNSON and S.J.POON, "Properties of Amorphous and Microcrystalline Superconductors", IEEE Trans. Mag., MAG-11 (1975) 189-91 (S).

291. W.L.JOHNSON and S.J.POON, "Superconductivity in Amorphous and Microcrystalline Transition-Metal Alloys", J. Appl. Phys., 46 (1975) 1787-92 (S).

292. W.L.JOHNSON and S.J.POON, "Enhanced Superconductivity and Lattice Instability in Nb-Rh Alloys", J. Appl. Phys., 46 (1975) 2325-6 (S).

293. W.L.JOHNSON, S.J.POON and P.DUWEZ, "Amorphous Superconducting Lanthanum-Gold Alloys Obtained by Liquid Quenching", Phys. Rev., B11 (1975) 150-4 (S).

294. N.S.KAZAMA and H.WATANABE, "Study of a Magnetic Phase Transition in Amorphous Ferromagnets with Polarized Neutrons", J. Phys. Soc. Japan, 39 (1975) 1411-2 (Mag).

295. M.KIKUCHI, H.FUJIMORI, Y.OBI and T.MASUMOTO, "New Amorphous Ferromagnets with Low Coercive Force", Japan. J. Appl. Phys., 14 (1975) 1077-8 (Mag).

296. H.KIMURA and T.MASUMOTO, "Fracture Toughness of Amorphous Metals", Scripta Met., 9 (1975) 211-21 (M).

297. R.KREPSKI, K.SWYLER, H.R.CARLETON and H.HERMAN, "Liquid Quenching of Laser-Melted Oxides", J. Mater. Sci., 10 (1975) 1452-4 (QT).

298. K.D.KRISHNANAND and R.W.CAHN, "Recovery from Plastic Deformation in a Ni/Nb Alloy Glass", Scripta Met., 9 (1975) 1259-61 (M).

299. D.KUNSTELJ and A.BONEFAČIĆ, "Phase Transformations in an Annealed, Liquisol-Quenched Al-3.6 at.%Ni Alloy", in Microstructural Science, eds. P.M.French, R.J.Gray and J.L.McCall, (American Elsevier, New York, 1975) Vol. 3A, pp. 207-15 (AB).

300. M.LASOCKA and H.MATYJA, "Thermodynamics of Non-Crystalline Splat-Cooled Foils", Mater. Sci. & Engg., 19 (1975) 211-6 (PM).

301. M.LASOCKA, P.G.ZIELINSKI and H.MATYJA, "The Effect of the Properties of Alloying Components on the Vitrification Ability as Exemplified by Some Tellurium Alloys", Scripta Met., 9 (1975) 873-8 (PM).

302. H.J.LEAMY, S.D.FERRIS, G.NORMAN, D.C.JOY, R.C.SHERWOOD, E.M.GYORGY and H.S.CHEN, "Ferromangetic Domain Structure of Metallic Glasses", Appl. Phys. Letters, 26 (1975) 259-60 (Mag).

303. D.LESUEUR, "Amorphisation of a Pd-Si Alloy by Fission Fragment Irradiation", Radiation Effects, 24 (1975) 101-10 (in French) (AS)(see Phys. Abstr., 78(1975) 37276).

304. J.LOGAN, "The Structure of an Amorphous Superconductor, Lanthanum-Gold", Scripta Met., 9 (1975) 379-82 (AQS).

305. J.LOGAN, "The Radial Distribution Function of Amorphous Iron-Phosphorus Alloys", Phys. Stat. Sol., (a) 32 (1975) 361-8 (AQS).

306. F.E.LUBORSKY, J.J.BECKER and R.O.McCARY, "Magnetic Annealing of Amorphous Alloys", IEEE Trans. Mag., MAG-11 (1975) 1644-9 (Mag).

307. A.B.LYSENKO, V.I.TKACH and A.A.YAKUNIN, "Formation and Decomposition of Non-Crystalline Phases in La-Al and La-Mn Alloys Quenched from a Melt", V Sb., Struktura Zhidkosti i Fazovye Perekhody, Dnepropetrovsk, No. 3 (1975) 54-60 (in Russian) (AS, AB)(see Chem. Abstr., 84(1976) 125459x).

308. F.MACHIZAUD, F.A.KUHNAST and J.FLECHON, "Local Order and D.T.A. Studies of Ni-P", J. Physique, 36 (1975) C2-83-8 (in French) (VD; AQS).

309. M.L.MALHOTRA, and K.BASU, "Microstructure in Gold-Containing Ag_3Sn Splat-Cooled Ribbons", Metallography, 8 (1975) 515-23 (AQS).

310. G.MARCHAL, Ph.MANGIN and C.JANOT, "Crystal Growth and Precipitation in Thin Films of Amorphous Fe-Au Alloys", Phil. Mag., 32 (1975) 1007-21 (VD; AB).

311. T.MASUMOTO, "Recent Developments of Amorphous Metals", Kinzoku, 45 , No. 6 (1975) 59-64 (in Japanese) (R)(see Chem. Abstr., 84 (1976) 63816a).

312. T.MASUMOTO, K.HASHIMOTO and H.FUJIMORI, "Recent Development of Research on Amorphous Metals", Sci. Rep. RITU, A25 (1975) 232-44 (R).

313. T.MASUMOTO and H.KIMURA, "Tensile Properties of Iron-base Amorphous Alloy (Fe-P-C) Quenched from Liquid", J. Japan Inst. Metals, 39 (1975) 133-41 (in Japanese); Sci. Rep. RITU, A25 (1975) 200-5 (M)(see Met. Abstr., 8(1975) 31-3307).

314. T.MASUMOTO and H.KIMURA, "Crystallization Process of Iron-base Amorphous Alloy (Fe-P-C) Quenched from Liquid", J. Japan Inst. Metals, 39 (1975) 273-80 (in Japanese); Sci. Rep. RITU, A25 (1975) 216-31 (AB)(see Met. Abstr., 8(1975) 12-1447).

315. T.MASUMOTO and R.MADDIN, "Structural Stability and Mechanical Properties of Amorphous Metals - A Review", Mater. Sci. & Engg., 19 (1975) 1-24 (R).

316. H.MATYJA, K.C.RUSSELL, B.C.GIESSEN and N.J.GRANT, "Precipitation of Silicon from Splat-Cooled Al-Si Alloys", Met. Trans., A6 (1975) 2249-52 (AB).

317. I.MIKI, H.KOSUGE and K.NAGAHAMA, "Supersaturation and Decomposition of Al-Fe Alloys During Solidification", J. Japan Inst. Light Metals, 25 (1975) 1-9 (in Japanese) (AB)(see Met. Abstr., 8(1975) 21-0528).

318. I.S.MIROSHNICHENKO, "Formation of Metastable Phases during Solidification of Rapidly Cooled Melts", V Sb., Rost. Kristallov, Erevan, No. 11 (1975) 337-48 (in Russian)(see Chem. Abstr., 84 (1976) 109590c).

319. I.S.MIROSHNICHENKO, G.P.BREKHARYA and B.L.TKACHENKO, "Determining the Rate of Cooling on Quenching from the Liquid State", Fiz. Khim. Obrab. Mater., No. 6 (1975) 53-6 (PM)(see Met. Abstr., 9(1976) 12-0712).

320. T.MIZOGUCHI, K.YAMAUCHI and N.NIMURA, "Neutron Diffraction Study of an Amorphous $Fe_{80}P_{13}C_7$ Alloy with ToF Technique", Japan. J. Appl. Phys., 14 (1975) 711-2 (AQS).

321. H.A.MOOK, D.PAN, J.D.AXE and L.PASSELL, "Magnetic Excitations in Amorphous Co_4P", AIP Conf. Proc., 24 (1975) 112 only (ED; Mag).

322. H.A.MOOK, N.WAKABAYASHI and D.PAN, "Magnetic Excitations in the Amorphous Ferromagnet Co_4P", Phys. Rev. Letters, 34 (1975) 1029-33 (ED; Mag).

323. P.MRAFKO and P.DUHAJ, "A Compromise Hard Sphere Model for Amorphous Metallic Alloys", J. Non-Cryst. Solids, 17 (1975) 143-5 (SMA).

324. S.R.NAGEL and J.TAUC, "Nearly-Free-Electron Approach to the Theory of Metallic Glass Alloys", Phys. Rev. Letters, 35 (1975) 380-3 (AS).

325. P.K.K.NAYAR, "Influence of High Cooling Rates on the Growth of the Product in Liquid-Solid Transformations", Proc. Symp. Phase Transformation and Phase Equilibria (Dept. of Atomic Energy, India, 1975) pp. 401-6 (AQS).

326. T.OGURA, K.FUKUSHIMA and T.MASUMOTO, "Propagation of Fatigue Cracks in Amorphous Metals", Scripta Met., 9 (1975) 979-83 (M).

327. T.OGURA, T.MASUMOTO and K.FUKUSHIMA, "Fatigue Fracture of Amorphous Pd-20 at.% Si Alloy", Scripta Met., 9 (1975) 109-13 (M).

328. R.C.O'HANDLEY, "Domain Wall Kinetics in Soft Ferromagnetic Metallic Glasses", J. Appl. Phys., 46 (1975) 4996-5001 (Mag).

329. C.A.PAMPILLO, "Flow and Fracture in Amorphous Alloys", J. Mater. Sci., 10 (1975) 1194-1227 (R).

330. D.E.POLK and D.S.BOUDREAUX, "Comment on the Interpretation of the First Peak of the RDF of Amorphous Metals", J. Physique, 36 (1975) C2-55-8 (AQS).

331. S.J.POON and W.L.JOHNSON, "Phenomenological Approach to High-Transition-Temperature Superconducting Alloys", Phys. Rev., B12 (1975) 4816-24 (S).

332. D.V.RATNAM, K.S.V.L.NARASIMHAN and B.C.GIESSEN, "Preparation and Characterization of Splat-Cooled R-Co Compounds", AIP Conf. Proc., 24 (1975) 687-8 (AQS).

333. A.REVCOLEVSCHI, "Application of Splat Cooling from the Liquid Phase to Oxides", Acta Technica Belgica-Metallurgie, 15 (1975) 169-75 (in French) (R, AS, MCS)(see Met. Abstr., 9 (1976) 11-0547).

334. A.REVCOLEVSCHI, A.ROUANET, F.SIBIEUDE and T.SUZUKI, "Techniques for Ultrafast Quenching (Splat Cooling) of Refractory Oxides", High Temp. High Pressures, 7 (1975) 209-14 (QT, R).

335. A.V.ROMANOVA and V.V.BUKHALENKO, "Crystallographic X-Ray Investigation of the Alloy Cu-9.2 at.% Ga 'Quenched' from the Liquid State", Fiz. Met. Metalloved., 40 (1975) 884-5 (in Russian); Phys. Met. Metallogr., 40, No. 4 (1975) 193-5 (AQS).

336. R.B.ROOF and R.O.ELLIOTT, "Evidence for the Existence of Faulting in a Splat-Cooled δ-Pu(Ti) Alloy", J. Mater. Sci., 10 (1975) 101-8 (AQS).

337. A.G.RUDENKO and L.D.CHUMAKOVA, "Effect of Very Rapid Quenching from the Melt on the Structure of Fe-Mn Alloys", Izv. Akad. Nauk SSSR, Met. No. 6 (1975) 129-33 (in Russian) (AQS) (see Met. Abstr., 9(1976) 12-0581).

338. M.RÜHLE, "New Dispersion-Strengthened Aluminium Alloys", Metall, 29 (1975) 237-47 (in German) (M).

339. J.F.SADOC, "Distribution of Magnetic Moments in Metallic Amorphous Alloys", J. Physique, 36 (1975) C2-75-8 (in French) (Mag).

340. I.R.SARE, "Two Non-Crystalline Phases in a Splat-Cooled Fe-1.9 wt.% Si-4.2 wt.% C Alloy", Scripta Met., 9 (1975) 607-9 (AS).

341. J.SCHNEIDER and H.WIESNER, "Magnetic Properties of Rapidly Quenched Ni-P Alloys", Phys. Stat. Sol., (a) 29 (1975) K25-9 (Mag).

342. J.SCHNEIDER and H.WIESNER, "Temperature Dependence of Magnetization in Fe-P Alloys Rapidly Quenched from the Liquid State", Phys. Stat. Sol., (a) 29 (1975) 151-60 (Mag).

343. J.SCHNEIDER and H.WIESNER, "Structure Transformation in Rapidly Quenched Iron-base Alloys", Phys. Stat. Sol., (a) 32 (1975) 565-9 (AB).

344. M.G.SCOTT, "The Role of Melt Superheat in Splat Quenching", J. Mater. Sci., 10 (1975) 269-73 (PM).

345. M.G.SCOTT, "X-Ray Diffraction and Differential Thermal Analysis of Splat-Quenched Gold-Germanium Alloy", Mater. Sci. & Engg., 18 (1975) 279-84 (AQS, AB).

346. M.G.SCOTT and J.A.LEAKE, "The Formation and Decomposition of an Al-17.3 at.% Cu Solid Solution", Acta Met., 23 (1975) 503-11 (TSSE, AB).

347. SHAMSUDDIN, P.RAMACHANDRARAO, S.MISRA and T.R.ANANTHARAMAN, "Thermodynamic and Constitutional Studies of the PbTe-GeTe System", J. Mater. Sci., 10 (1975) 1849-55 (T, AB).

348. S.K.SHARMA, H.P.GESERICH and W.A.THEINER, "Amorphous-to-Crystalline Transformation in Evaporated Fe-Si Films", Phys. Stat. Sol., (a) 32 (1975) 467-74 (VD; AB).

349. R.C.SHERWOOD, E.M.GYORGY, H.S.CHEN, S.D.FERRIS, G.NORMAN and H.J.LEAMY, "Ferromagnetic Behavior of Metallic Glasses", AIP Conf. Proc., 24 (1975) 745-6 (Mag).

350. P.H.SHINGU and R.OZAKI, "Solidification Rate in Rapid Conduction Cooling", Met. Trans., A6 (1975) 33-7 (PM).

351. A.W.SIMPSON and W.G.CLEMENTS, "Magnetostriction in Some Amorphous Fe-Co-Ni-P Alloys", IEEE Trans. Mag., MAG-11 (1975) 1338-40 (CD; Mag).

352. F.SPAEPEN, "On the Fracture Morphology of Metallic Glasses", Acta Met., 23 (1975) 615-20 (M).

353. S.TAKAYAMA and R.MADDIN, "Resistivity Changes by Deformation of Ni-Pd-P Amorphous Alloys", Scripta Met., 9 (1975) 343-7 (E).

354. S.TAKAYAMA and R.MADDIN, "Rolling and Bending Deformation of Ni-Pd-P Metal Glasses", Acta Met., 23 (1975) 943-52 (M).

355. S.TAKAYAMA and R.MADDIN, "Fracture of Amorphous Ni-Pd-P Alloys", Phil. Mag., 32 (1975) 457-70 (M).

356. G.L.TANGONAN, "Electrical Resistivity of Amorphous Pd-Cu-P Alloys", Phys. Letters, A54 (1975) 307-8 (E).

357. K.TOGANO and K.TACHIKAWA, "Structure and Superconductivity of Metastable Phases in Liquid-Quenched Zr-Rh Alloys", J. Appl. Phys., 46 (1975) 3609-13 (S).

358. K.TOGANO and K.TACHIKAWA, "Upper Critical Fields of Non-Crystalline $Zr_{77}Rh_{23}$ Alloy and Intermetallic Zr_2Rh Compound", Phys. Letters, A54 (1975) 205-6 (S).

359. N.TSUYA, K.I.ARAI, Y.SHIRAGA and T.MASUMOTO, "Observation of the Magnetostriction in Ferromagnetic Amorphous Thin Ribbons", Phys. Letters, A51 (1975) 121-2 (Mag).

360. N.TSUYA, K.I.ARAI, Y.SHIRAGA, M.YAMADA and T.MASUMOTO, "Magnetostriction of Amorphous $Fe_{0.80}P_{0.13}C_{0.07}$ Ribbon", Phys. Stat. Sol., (a) 31 (1975) 557-61 (Mag).

361. D.TURNBULL, "Structure and Properties of Glassy Metals", J. Electron. Mater., 4 (1975) 771-81 (R)(see Met. Abstr., 9 (1976) 12-0368).

362. N.I.VARICH, A.A.YAKUNIN and A.B.LYSENKO, "The Formation and Decomposition of a Non-Crystalline Phase in La-25 at.% Ag Alloy Quenched from the Liquid State", Fiz. Tverd. Tela., 17 (1975) 1824-6 (in Russian); Soviet Phys.-Solid State, 17 (1975) 1194-5 (AS, AB)(see Met. Abstr., 8(1975) 11-0823).

363. J.M.VITEK, "Characteristics of Electron-Transparent Regions in Unthinned, Rapidly Quenched Foils", J. Mater. Sci., 10 (1975) 1828-9 (PM).

364. J.M.VITEK and N.J.GRANT, "Mechanical Properties of an Amorphous Cu-Zr Alloy", Met. Trnas., A6 (1975) 1472-3 (M).

365. J.M.VITEK, J.B.VANDER SANDE and N.J.GRANT, "Crystallization of an Amorphous Cu-Zr Alloy", Acta Met., 23 (1975) 165-76 (AB).

366. Y.WASEDA, "The Structure of Amorphous Metallic Alloys", Solid State Phys., 10 (1975) 459-74 (in Japanese) (R)(see Phys. Abstr., 79 (1976) 25956).

367. Y.WASEDA, "Oxide Glasses and Metallic Glasses", Kinzoku, 45, No. 6 (1975) 51-5 (in Japanese) (R).

368. Y.WASEDA and T.MASUMOTO, "An X-Ray Diffraction Study of the Structure of Amorphous $Cu_{57}Zr_{43}$ Alloy", Z. Physik, B21 (1975) 235-40 (AQS).

369. Y.WASEDA and T.MASUMOTO, "Structure of Amorphous $Fe_{80}P_{13}C_7$ by X-Ray Diffraction", Z. Physik, B22 (1975) 121-6 (AQS).

370. Y.WASEDA and T.MASUMOTO, "Structural Study of an Amorphous 80:20 Palladium-Silicon Alloy by X-Ray Fourier Analysis", Phys. Stat. Sol., (a) 31 (1975) 477-82 (AQS).

371. T.WATANABE and Y.TANABE, "Amorphous Nickel-Boron Alloy Film Deposited by Electroless Deposition", J. Japan Inst. Metals, 39 (1975) 831-6 (in Japanese) (CD; AS, AQS)(see Phys. Abstr., 79(1976) 13885).

372. H.WIESNER and J.SHNEIDER, "Temperature and Field Dependence of Magnetization of Rapidly Quenched Co-P Alloys", Phys. Stat. Sol., (a) 29 (1975) 503-9 (Mag).

373. C.A.WILLIAMS, and H.JONES, "The Effect of Melt Superheat and Impact Velocity on Splat Thickness", Mater. Sci. & Engg., 19 (1975) 293-7 (PM).

374. J.V.WOOD and I.R.SARE, "Experimental Observations of Crystal Growth in Alloys Rapidly Quenched from the Liquid State", Met. Trans., A6 (1975) 2153-5 (AQS).

375. S.YAJIMA, K.OKAMURA and T.SHISHIDO, "Amorphous Phase in Yttrium-Cobalt-Boron System", Chem. Letters, No. 12 (1975) 1227-30 (AS).

376. K.YAMADA, Y.ISHIKAWA, Y.ENDOH and T.MASUMOTO, "The Magnetic Phase Transition of an Amorphous Fe-P-C and its Alloys Containing Ni and Cr", Solid State Commun., 16 (1975) 1335-8 (Mag).

377. K.YAMAUCHI and T.MIZOGUCHI, "The Magnetic Moment of Amorphous Metal-Metalloid Alloys", J. Phys. Soc. Japan, 39 (1975) 541-2 (Mag).

378. A.ZENTKO, P.DUHAJ, L.POTOCKÝ, T.TIMA and J.BÁNSKY, "Low Field Magnetic Susceptibility of Amorphous $Co_xPd_{80-x}Si_{20}$ and Fe_x-$Pd_{80-x}Si_{20}$ Alloys", Phys. Stat. Sol., (a) 31 (1975) K41-2 (Mag).

379. E.ZIAI and Y.B.KIM, "Mechanism for Coercivity in Splat-Cooled RCo_5 Compounds", AIP Conf. Proc., 24 (1975) 685-6 (Mag).

1976

380. R.ALBEN, G.S.CARGILL III and J.WENZEL, "Anisotropy of Strcutural Models for Amorphous Materials", Phys. Rev., B13 (1976) 835-42 (SMA).

381. L.A.ALEKSEEV, V.T.BORISOV, A.I.DUKHIN, A.M.DURACHENKO, R.B.LEVI and A.F.PROKOSHIN, "Study of Magnetic Anisotropy in Amorphous Fe-P-C Alloys by the Nuclear Gamma Resonance Method", Fiz. Met. Metalloved., 42 (1976) 887-9 (in Russian); Phys. Met. Metallogr., 42, No. 4 (1976) 190-2 (Mag).

382. A.AMAMOU, "Magnetism and Local Environmental Model in $(Ni_{1-c}Co_c)_{0.78}P_{0.14}B_{0.08}$ Amorphous Alloys", IEEE Trans. Mag., MAG-12 (1976) 948-50 (Mag).

383. A.AMAMOU and J.DURAND, "Giant Moments and Short Range Order in Amorphous Ni-P-B Alloys", Commun. Phys., 1 (1976) 191-7 (Mag) (see Phys. Abstr. 80(1977) 4499).

384. P.ANDONOV, "Comparison of Amorphous and Liquid Pd-Si Alloys", J. Non-Cryst. Solids, 22 (1976) 145-58 (SMA).

385. R.ANDREWS, W.A.GRANT, P.J.GRUNDY, J.S.WILLIAMS and L.T.CHADDERTON, "Non-Crystallinity of Ion-Irradiated Metals", Nature, 262 (1976) 380-1 (II; AS).

386. K.I.ARAI, N.TSUYA, M.YAMADA and T.MASUMOTO, "Giant ΔE Effect and Magnetomechanical Coupling Factor in Amorphous $Fe_{80}P_{13}C_7$ Ribbons", IEEE Trans. Mag., MAG 12 (1976) 936-8 (Mag).

387. K.I.ARAI, N.TSUYA, M.YAMADA amd T.MASUMOTO, "Zero Magnetostriction and Extremely Low Residual Magnetic Loss in Fe-Co Amorphous Ribbons", IEEE Trans. Mag., MAG 12 (1976) 939-41 (Mag).

388. Y.ARATA, F.MATSUDA and K.NAKATA, "Effect of Solidification Rate on Solidification Structure in Weld Metal", Trans. JWRI, 5 No. 1 (1976) 47-52 (AQS)(see Met. Abstr., 9(1976) 12-1171).

389. P.B.ARONSON, "The Promise of Quick-Quench Materials", Mach. Des., 48 No. 28 (1976) 20-4 (R).

390. K.ASAMI, K.HASHIMOTO, T.MASUMOTO and S.SHIMODAIRA, "ESCA Study of the Passive Film on an Extremely Corrosion-Resistant Amorphous Iron Alloy", Corrosion Sci., 16 (1976) 909-14 (C).

391. D.G.AST and D.J.KRENITSKY, "Pressure Dependence and Tensile Stress Dependence of the Resistivity of an Fe-Ni Base Metallic Glass", Scripta Met., 10 (1976) 247-9 (E).

392. M.AUSLOOS, "Thermodynamics of Amorphous Magnetic Metals and Alloys at High Temperature", AIP Conf. Proc., 34 (1976) 304-6 (Mag).

393. J.D.AXE, "Magnetic Excitations in Amorphous Metallic Ferromagnets", AIP Conf. Proc., 29 (1976) 146 only (Mag).

394. E.BABIĆ and G.GRÜNER, "Interaction Effects in Dilute Alloys", Physica, 84 B (1976) 37-9 (E).

395. E.BABIĆ and G.GRÜNER, "Interactions in Kondo Alloys", Fizika, 8, Suppl. (1976) 219 (E).

396. E.BABIĆ, R.KRSNIK and M.OČKO, "The Low-Temperature Resistivity of AlV and AlTi Alloys", J. Phys. F: Metal Phys., 6 (1976) 73-83 (E).

397. F.BAUMANN, "Metastable Phases Frozen in by Quenching Condensation", Thin Solid Films, 34 (1976) 87-90 (VD; R).

398. J.J.BECKER, "Domain Observations in an Amorphous Fe-Ni Alloy", AIP Conf. Proc., 29 (1976) 204-6 (Mag).

399. G.BELLESSA, M.CAGNON, J.F.SADOC, P.DOUSSINEAU and A.LEVELUT, "Sound Propagation in Amorphous Metallic Ni-P", J. Physique Letters, 37 (1976) 291-3 (ED; PP).

400. G.BERGMANN, "Amorphous Metals and Their Superconductivity", Phys. Rep., 27 (1976) 159-85 (VD; S, R).

401. B.S.BERRY and W.C.PRITCHET, "Magnetoelasticity and Internal Friction of an Amorphous Ferromagnetic Alloy", J. Appl. Phys., 47 (1976) 3295-301 (M).

402. B.S.BERRY and W.C.PRITCHET, "Magnetoelastic Phenomena in Amorphous Alloys", AIP Conf. Proc., 34 (1976) 292-7 (Mag).

403. J.W.M.BIESTERBOS and A.G.DIRKS, "Amorphous Metallic Systems", Polytech. Tijdschr. Werktuigbouw, 31 (1976) 145-50 (in Dutch) (R)(see Phys. Abstr., 79(1976) 49648).

404. C.BIRAC and D.LESUEUR, "Atomic Diffusion of Lithium in the Amorphous Metallic Alloy Pd-Si", Phys. Stat. Sol., (a) 36 (1976) 247-51 (in French) (PP)(see Phys. Abstr., 79(1976) 79601).

405. J.BLÉTRY, "Partial Structure Factors in Multicomponent Liquid or Amorphous Alloys", Z. Naturforsch., A31 (1976) 960-6 (AQS).

406. V.T.BORISOV, A.I.DUKHIN and A.M.DURACHENKO, "Study of Formation Conditions of Iron-Based Amorphous Alloys During Quenching from a Liquid", Nauch. Soobshch. II Vses. Konf. po Stroeniyu i Svoitsvam Met. i Shlak Rasplavov, Chapter 2 (1976) pp. 107-8 (PM)(see Chem. Abstr., 86(1977) 174897p).

407. P.G.BOSWELL and G.A.CHADWICK, "Cellular Growth in a $Pd_{781}Cu_{55}$-Si_{164} Amorphous Alloy", Scripta Met., 10 (1976) 509-13 (AQS).

408. P.G.BOSWELL and G.A.CHADWICK, "A Casting Device for Producing Rapidly Solidified Metallic Strips", J. Phys. E: Sci. Instr., 9 (1976) 523-6 (QT).

409. P.G.BOSWELL and C.A.CHADWICK, "The Formation and Crystallization of An Amorphous Phase in an Fe-C Alloy", J. Mater. Sci., 11 (1976) 2287-96 (AS, AB).

410. E.M.BREINAN, B.H.KEAR and C.M.BANAS,"Processing Materials with Lasers", Phys. Today, 29, No. 11 (1976) 44-8 (QT).

411. C.L.BRIANT, "On a Structural Model for Amorphous Metals. Implications from Microcluster Studies", Faraday Disc. Chem. Soc., No. 61 (1976) 25-33 (SMA)(see Chem. Abstr., 87(1977) 72002w).

412. H.A.BROOKS, "Magnetostriction vs Cobalt Content in Amorphous Alloys of Fe-Co-P-B-Al", J. Appl. Phys., 47 (1976) 344-5 (Mag).

413. H.A.BROOKS and H.S.CHEN, "Magnetostriction as a Function of Glass Formers in Amorphous Alloys of Fe-P-B-Al", AIP Conf. Proc., 29 (1976) 208 only (Mag).

414. H.J.BUNGE, "Textures and Metastable Phases in Taylor Wires", Z. Metallkde., 67 (1976) 720-8 (MCS).

415. P.BUTVIN and P.DUHAJ, "The Hall Effect in PdSi-based Amorphous Alloys containing Co I.Nonmagnetic Alloys", Czech. J. Phys., B 26 (1976) 208-14 (Mag)(see Phys. Abstr., 79(1976) 31106).

416. P.BUTVIN and P.DUHAJ, "The Hall Effect in PdSi-based Amorphous Alloys Containing Co II.Ferromagnetic-Like Alloys", Czech. J. Phys., B 26 (1976) 469-76 (Mag)(see Phys. Abstr., 79(1976) 45308).

417. R.W.CAHN, "Rapid Quenching: Second Harvest", Nature, 259 (1976) 271-2.

418. B.CANTOR and R.W.CAHN, "Vapour-Quenched Silver-Copper Alloys", Scripta Met., 10 (1976) 381-2 (VD; AQS).

419. B.CANTOR and R.W.CAHN, "Precipitation of Equilibrium Phases in Vapour-Quenched Al-Ni, Al-Cu and Al-Fe Alloys", J. Mater. Sci., 11 (1976) 1066-76 (VD; AB, TSSE).

420. B.CANTOR and R.W.CAHN, "Metastable Alloy Phases by Co-Sputtering", Acta Met., 24 (1976) 845-52 (Sp; AQS).

421. G.S.CARGILL III, "Recurring Themes in the Structure of Glassy Solids", Annals N.Y. Acad. Sci., 279 (1976) 208-22 (SMA)(see Chem. Abstr., 86(1977) 76980r).

422. G.S.CARGILL III and S.KIRKPATRICK, "Structural Ordering in Some Amorphous Metals and Alloys", AIP Conf. Proc., 31 (1976) 339-52 (AQS).

423. K.CHATTOPADHYAY and P.RAMACHANDRARAO, "Multiply Twinned Particles in an Al-Ni Vapor Deposit", J. Cryst. Growth, 36 (1976) 355-7 (VD; AQS).

424. L.A.CHEKANOVA, R.S.ISKHAKOV, G.I.FISH, R.G.KHLEBEPROS, N.S. CHISTYAKOV and Y.N.AGARTANOVA, "Spin Wave Resonance and Structural Transformations in Amorphous Films", Fiz. Met. Metalloved., 41 (1976) 536-41 (in Russian); Phys. Met. Metallogr., 41, No. 3 (1976) 72-7 (VD; AB).

425. H.S.CHEN, "Stored Energy in a Cold-Rolled Metallic Glass" Appl. Phys. Letters, 29 (1976) 328-30 (M).

426. H.S.CHEN, "Positron Lifetime Study on the Structure of an Electron-Irradiated Metallic Glass", Phys. Stat. Sol., (a) 34 (1976) K127-9 (AQS).

427. H.S.CHEN, "Correlation between the Thermal Stability and Activation Energy of Crystallization in Metallic Glasses", Appl. Phys. Letters, 29 (1976) 12-4 (AB).

428. H.S.CHEN, "Entropy Model for Flow Behavior in Metallic Glasses", J. Non-Cryst. Solids, 22 (1976) 135-43 (M).

429. H.S.CHEN, "Metallic Glasses", Mater. Sci. & Engg., 25 (1976) 59-69 (R).

430. H.S.CHEN, "Glass Temperature and Stability of Fe, Co, Ni and Pd-Based Glasses: Effect of Configurational Entropy and Crystalline Symmetry", Acta Met., 24 (1976) 153-8 (T, AS).

431. H.S.CHEN, "Ductile-Brittle Transition in Metallic Glasses", Mater. Sci. & Engg., 26 (1976) 79-82 (M).

432. H.S.CHEN and E.COLEMAN, "Structure Relaxation Spectrum of Metallic Glasses", Appl. Phys. Letters, 28 (1976) 245-7 (AQS).

433. H.S.CHEN and C.E.MILLER, "Centrifugal Spinning of Metallic Glass Filaments", Mater. Res. Bull., 11 (1976) 49-54 (QT).

434. H.S.CHEN, R.C.SHERWOOD, H.J.LEAMY and E.M.GYORGY, "The Effect of Structural Relaxation on the Curie Temperature of Fe-based Metallic Glasses", IEEE Trans. Mag., MAG-12 (1976) 933-5 (Mag).

435. M.C.CHI and R.ALBEN, "Hysteresis Curves for a Random Packing Model of an Amorphous Magnet with Uniaxial Anisotropy", AIP Conf. Proc., 34 (1976) 316-8 (Mag).

436. G.C.CHI and G.S.CARGILL III, "Perpendicular Anisotropy of Amorphous, Electrodeposited Co-P Alloy Films", AIP Conf. Proc., 29 (1976) 147-8 (ED; Mag).

437. G.C.CHI and G.S.CARGILL III, "Radial Distribution Functions for Amorphous Co-P Alloys", AIP Conf. Proc., 31 (1976) 359-65 (ED; AQS).

438. C.L.CHIEN and R.HASEGAWA, "Magnetic Properties of a Ferromagnetic Glass $Fe_{80}P_{16}B_1C_3$", IEEE Trans. Mag., MAG-12 (1976) 951-3 (Mag).

439. C.L.CHIEN and R.HASEGAWA, "Mössbauer Study of Amorphous Fe_{75}-$P_{15}C_{10}$", J. Physique, 37 (1976) C6-759-61 (Mag).

440. C.L.CHIEN and R.HASEGAWA, "Easy Axis of Amorphous Ferromagnet $Fe_{40}Ni_{40}P_{14}B_6$", J. Appl. Phys., 47 (1976) 2234-6 (Mag).

441. C.L.CHIEN and R.HASEGAWA, "Magnetic Properties of Amorphous $Fe_{40}Ni_{40}P_{14}B_6$", AIP Conf. Proc., 29 (1976) 214-6 (Mag).

442. C.L.CHIEN and R.HASEGAWA, "Spontaneous Magnetization of Ferromagnetic Glasses", AIP Conf. Proc., 31 (1976) 366-71 (Mag).

443. C.P.CHOU, "Correlation of Elastic and Magnetic Properties of Magnetic Glassy Alloys", Phys. Rev. Letters, 37 (1976) 1004-7 (M, Mag).

444. J.R.COOPER and M.MILJAK, "Single Impurity Behaviour and Interaction Effects in the Magnetic Susceptibility of AlMn and AlCr Alloys", J. Phys. F: Metal Phys., 6 (1976) 2151-64 (Mag).

445. P.J.COTE, "Electrical Resistivity of Amorphous Nickel-Phosphorus Alloys", Solid State Commun., 18 (1976) 1311-3 (E).

446. R.M.J.COTTERILL, "Glassy Metals", Amer. Sci., 64 (1976) 430-7 (R)(see Phys. Abstr., 80(1977) 1081).

447. P.R.COUCHMAN, C.L.REYNOLDS and R.M.J.COTTERILL, "Temperature Dependence of the Amorphous Specific Heat of Glasses and Amorphous Solids", Nature, 259 (1976) 108-9 (T).

448. P.R.COUCHMAN, C.L.REYNOLDS and R.M.J.COTTERILL, "low Temperature Specific Heat of Glasses and Amorphous Solids", Nature, 264 (1976) 534-5 (T).

449. J.P.COUTURES, E.ANTIC and P.CARO "A Novel Phase Obtained in the System $Nd_2O_3 \cdot Al_2O_3$ by Rapid Cooling from the Liquid State and its Comparison with $Nd_4Ga_2O_9$", Mater. Res. Bull., 11 (1976) 699-706 (in French)(MCS).

450. H.A.DAVIES, "The Formation of Metallic Glasses", Phys. & Chem. Glasses, 17 (1976) 159-73 (R).

451. H.A.DAVIES and J.B.HULL, "Comments on 'On the Presence of Amorphous Regions in Splat-Quenched Al-Cu Alloys", J. Mater. Sci., 11 (1976) 2149-51 (AQS)(see entry 614).

452. H.A.DAVIES and J.B.HULL, "The Formation, Structure and Crystllization of Non-Crystalline Nickel Produced by Splat Quenching", J. Mater. Sci., 11 (1976) 215-23 (AS, AB).

453. H.A.DAVIES and H.JONES, "New Materials by Rapid Quenching from the Melt", Metals and Materials, (June 1976) 44-5.

454. H.A.DAVIES and B.G.LEWIS, "Discussion of 'Solidification Rate in Rapid Conduction Cooling'", Met. Trans., A7 (1976) 310-2 (PM)(see entry 350).

455. H.A.DAVIES and B.G.LEWIS, "Metallic Glasses and the Avoidance of Crystallization in Metallic Melts", Faraday Disc. Chem. Soc., No. 61 (1976) 100-9 (AS, PM)

456. L.A.DAVIS, "Fatigue of Metallic Glasses", J. Mater. Sci., 11 (1976) 711-7 (M).

457. L.A.DAVIS, C.P.CHOU, L.E.TANNER and R.RAY, "Strength and Stiffnesses of Metallic Glasses", Scripta Met., 10 (1976) 937-40 (M).

458. L.A.DAVIS, R.RAY, C.P.CHOU and R.C.O'HANDLEY, "Mechanical and Thermal Properties of $Fe_{80}B_{20}$ Glass", Scripta Met., 10 (1976) 541-6 (M, T).

459. A.M.de GRAAF, "Comment on the Spin Wave Spectrum of Co_4P-Comparison with a Continuum Theory of Amorphous Ferromagnetism", AIP Conf. Proc., 29 (1976) 149 only (ED; Mag).

460. A.DENEUVILLE, P.GERARD and J.DEVENYI, "Effect of Annealing on an Amorphous Ge_xTe_{1-x} Matrix with Te Crystallites", J. Non-Cryst. Solids, 22 (1976) 77-88 (Sp, AB).

461. T.M.DEVINE and L.WELLS, "Anodic Polarization Behaviour of an Amorphous Metal ", Scripta Met., 10 (1976) 309-10 (C).

462. R.B.DIEGLE and J.E.SLATER, "Influence of Crystallinity on Corrosion Behavior of Ferrous Alloys", Corrosion, 32 (1976) 155-7 (C).

463. K.DOI, "Profile Analysis of Amorphous Haloes by Means of a Fourier Transformation, with Special Reference to the Structures of Amorphous Pt-C and Ni-P Alloys", J. Appl. Cryst., 9 (1976) 382-90 (VD; AQS).

464. P.DUHAJ, D.BARANČOK and A.ONDREJKA, "The Study of Transformation Kinetics of the Amorphous Pd-Si Alloy", J. Non-Cryst. Solids, 21 (1976) 411-28 (AB).

465. P.DUHAJ, J.SITEK, M.PREJSA and P.BUTVIN, "Electron Microscopy and Mössbauer Study of the Amorphous Metallic Alloys Pd_{80-x}-Fe_xSi_{20}", Phys. Stat. Sol., (a) 35 (1976) 223-33 (AB).

466. J.DURAND, "Concentration Dependence of the Magnetic Properties in Amorphous Fe-P-B Alloys", IEEE Trans. Mag., MAG-12 (1976) 945-7 (Mag).

467. J.L.DURAND and M.F.LAPIERRE, "NMR of Co^{59} in Amorphous Co-P Alloys", J. Phys. F: Metal Phys., 6 (1976) 1185-92 (ED; Mag).

468. P.DUWEZ, "Structure and Properties of Glassy Metals", Annual Rev. Mater. Sci., 6 (1976) 83-117 (R).

469. T.EGAMI and P.J.FLANDERS, "Temperature Dependence of 'Magnetic Anisotropy' in Amorphous Alloys", AIP Conf. Proc., 29 (1976) 220-1 (Mag).

470. R.O.ELLIOTT and B.C.GIESSEN, "Splat Cooling of Some Actinide Alloys" in 'Plutonium and Other Actinides', eds. H.Bland and R.Lindner (North-Holland, Amsterdam, 1976) pp. 47-60 (TSSE, MCS)(see Phys. Abstr., 80(1977) 21134).

471. H.FAHLENBRACH, "Amorphous Metallic Materials and their Future Importance in Magnetism", Metall, 30 (1976) 1162-4 (in German) (Mag, R)(see Chem. Abstr., 86(1977) 109942k).

472. H.FAHLENBRACH, "Magnetically Soft Glass - A New Material", Elektrotechnik, 58 (1976) 14-7 (in German) (R)(see Phys. Abstr., 79(1976) 61987).

473. R.FERRO, R.MARAZZA, R.CAPELLI, G.RAMBALDI and F.BAFFI, "Some Metastable Alloys in the Silver-Bismuth System", J. Less-Common Metals, 46 (1976) 45-50 (MCS).

474. E.FIGUEROA, L.LUNDGREN, O.BECKMAN and S.M.BHAGAT, "The Amorphous Magnetization of Amorphous Metglas 2826A", Solid State Commun., 20 (1976) 961-4 (Mag).

475. M.FISCHER, H.-J.GÜNTHERODT, E.HAUSER, H.U.KÜNZI, M.LIARD and R.MÜLLER, "Electronic Structure of Amorphous Transition Metal Alloys", Phys. Letters, A55 (1976) 423-5 (AQS).

476. V.FRANETOVIĆ, A.KIRIN and A.BONEFAČIĆ, "Microstructure Investigations of Ag-In Alloy Rapidly Quenched from the Melt", Fizika, 8 (1976) 188-90 (AQS)(see Phys. Abstr., 80(1977) 59768).

477. R.L.FREED, "A TEM Study of the Crystallization of Two Non-Crystalline Cu-Zr Alloys", Proc. Annual Meet. EMSA, 34 (1976) 384-5 (AB)(see Chem. Abstr., 86(1977)59074f).

478. H.FUJIMORI, K.I.ARAI, H.SHIRAE, H.SAITO, T.MASUMOTO and N.TSUYA, "Magnetostriction of Iron-Cobalt Amorphous Alloys", Japan. J. Appl. Phys., 15 (1976) 705-6 (Mag).

479. H.FUJIMORI and T.MASUMOTO, "Magnetic Properties of an Fe-13P-7 C Amorphous Ferromagnet; The Effects of Stress, Stress Annealing and Magnetic-Field Annealing", Trans. Japan Inst. Metals, 17 (1976) 175-80 (Mag).

480. T.FUKUNAGA, M.MISAWA and K.SUZUKI, "T-O-F Pulsed Neutron Diffraction of Amorphous and Liquid $Pd_{0.8}Si_{0.2}$ Alloy", Res. Rep. Lab. Nucl. Sci. Tohoku Univ., 9 (1976) 96-102 (SMA).

481. N.FUNAKOSHI, T.KANAMORI and T.MANABE, "The Compositional Dependence of the Crystallization Temperature in Amorphous Co-Si-B Alloys", Japan. J. Appl. Phys., 15 (1976) 1833-4 (T).

482. J.P.GASPARD, "Density of States and Cohesive Energy of Amorphous Transition Metals", AIP Conf. Proc., 31 (1976) 372-7 (ED; AQS, PP).

483. E.GIRT, Z.GATALO, A.KURSUMOVIČ and B.LEONTIĆ, "Vacancy Transformations in Aluminum", Fizika, 8 (1976) 196-8 (AB)(see Phys. Abstr., 80(1977) 59770).

484. J.M.GORDON, J.H.GIBBS and P.D.FLEMING, "The Hard Sphere 'Glass Transition'", J. Chem. Phys., 65 (1976) 2771-8 (T).

485. T.GOTO, "Production of Metallic Fiber by the Method of Melt Spinning", Bull. Japan Inst. Metals, 15 (1976) 633-8 (in Japanese) (QT).

486. T.GOTO, H.MURAKAMI and M.NAGANO, "Quantitative Deformation Textures of Copper Filament Manufactured by the Glass Conjugated Melt Spinning", Sen-i Gakkaishi, 32 (1976) 61-6 (in Japanese) (M).

487. J.F.GRACZYK and P.CHAUDHARI, "Scattering Properties of Dense Random Packings of Hard Spheres", Phys. Stat. Sol., (b) 75 (1976) 593-606 (SMA).

488. C.D.GRAHAM, Jr., T.EGAMI, R.S.WILLIAMS and Y.TAKEI, "Annealing Effects in Amorphous Magnetic Alloys", AIP Conf. Proc., 29 (1976) 218-9 (Mag).

489. V.N.GUDZENKO and A.F.POLESYA, "Structure of Ni-V and Co-Cr Alloys Rapidly Cooled from the Melt", Fiz. Met. Metalloved., 41 (1976) 1106-9 (in Russian); Phys. Met. Metallogr., 41 No. 5 (1976) 189-93 (MCS).

490. E.M.GYORGY, H.J.LEAMY, R.C.SHERWOOD and H.S.CHEN, "Amorphous Alloys as Soft Magnetic Materials-II", AIP Conf. Proc., 29 (1976) 198-203 (Mag, R).

491. F.HABEREY, "Amorphous Magnetic Materials", Elektro-Anzeiger, 29 (1976) 537-9 (in German) (Mag)(see Phys. Abstr., 80(1977) 16700).

492. D.G.HALL and J.S.FAULKNER, "Spectral Density Functions for Amorphous Solids", in Proc. Conf. Neutron Scattering (U.S. Dept. Commerce, 1976) pp. 800-6 (see Phys. Abst. 82(1979)47269).

493. A.HAMZIĆ and E.BABIĆ, "The Dependence of the Magnetoresistance of AlMn Alloys on Magnetic Field and Impurity Concentration", Fizika, 8 (1976) 227-8 (Mag)(see Phys. Abstr., 80(1977) 62896).

494. R.HARRIS and D.ZOBIN, "A Phase Diagram for Amorphous Magnetic Alloys", AIP Conf. Proc., 29 (1976) 156-8 (Mag).

495. R.HASEGAWA, "Magnetization, Anisotropy and Coercivity of a Glassy Metallic Alloys", AIP Conf. Proc., 29 (1976) 216-7 (Mag).

496. R.HASEGAWA and C.L.CHIEN, "Mössbauer and its R.F. sideband Effects in Iron-Rich Glassy Alloys", Solid State Commun., 18 (1976) 913-6 (Mag).

497. R.HASEGAWA, W.A.HINES, L.T.KABACOFF and P.DUWEZ, "Nuclear Magnetic Resonance Study of ^{31}P and ^{195}Pt in the Metallic Glass System $(Ni_xPt_{1-x})_{75}P_{25}$", Solid State Commun., 20 (1976) 1035-8 (Mag).

498. R.HASEGAWA, R.C.O'HANDLEY and L.I.MENDELSOHN, "Advances in Ferromagnetic Metallic Glasses", AIP Conf. Proc., 34 (1976) 298-303 (Mag).

499. R.HASEGAWA, R.C.O'HANDLEY, L.E.TANNER, R.RAY and S.KAVESH, "Magnetization, Magnetic Anisotropy and Domain Patterns of $Fe_{80}B_{20}$ Glass", Appl. Phys. Letters, 29 (1976) 219-21 (Mag).

500. K.HASHIMOTO, T.MASUMOTO and S.SHIMODAIRA, "Passivity of Extremely Corrosion Resistive Iron Alloys", in Passivity and its Breakdown on Iron and Iron Base Alloys, eds. R.W.Staehle and H.Okada (National Association of Corrosion Engineers, Houston, Texas, 1976) pp. 34-7 (C).

501. K.HASHIMOTO, K.OSADA, T.MASUMOTO and S.SHIMODAIRA, "Characteristics of Passivity of Extremely Corrosion-Resistant Amorphous Iron Alloys", Corrosion Sci., 16 (1976) 71-6 (C).

502. M.HOARE, "Stability and Local Order in Simple Amorphous Packings", Annals N.Y. Acad. Sci., 279 (1976) 186-207 (R, SMA) (see Chem. Abstr., 86(1977) 36453q).

503. H.HUNAKUBO and M.AKAIKE, "Electrical Resistance Change of Amorphous Alloys Under Tensile Stress", Kotai Butsuri, B1 (1976) 223-6 (in Japanese) (E)(see Chem. Abstr., 86(1977) 49453q).

504. O.HUNDERI, "Optical Properties of Disordered and Amorphous Metal Films", Thin Solid Films, 37 (1976) 275-80 (PP).

505. T.IKEDA and S.NISHI, "Phase Decomposition of a Supersaturated Al-3%Mn Solid Solution Solidified Rapidly", J. Japan Inst. Metals, 40 (1976) 571-8 (in Japanese) (AB).

506. T.IKEDA and S.NISHI, "Kinetics of the Precipitation Process in a Supersaturated Al-3% Mn Solid Solution Solidified Rapidly", J. Japan. Inst. Metals, 40 (1976) 578-84 (in Japanese) (AB).

507. T.IKEDA, S.NISHI and H.SEO, "X-Ray Diffraction Study on Structure of Al-1 to 5% Mn Alloys Solidified by Rapid Cooling", J. Japan Inst. Metals, 40 (1976) 584-92 (in Japanese) (AQS).

508. Y.INOKUTI and B.CANTOR, "Splat-Quenched Fe-Ni Alloys", Scripta Met., 10 (1976) 655-9 (AQS).

509. W.L.JOHNSON and C.C.TSUEI, "Effect of Thermodynamic Fluctuations on the Shape and Width of the Resistive Transition in Three-Dimensional Amorphous Superconducting Alloys", Phys. Rev., B13 (1976) 4827-33 (S).

510. A.KAWASHIMA, K.HASHIMOTO and T.MASUMOTO, "Stress Corrosion Cracking of Amorphous Iron-Base Alloys", Corrosion Sci., 16 (1976) 935-43 (C).

511. N.KAZAMA, M.KAMEDA and T.MASUMOTO, "Magnetic Properties of Fe and Co-Based Amorphous Alloys", AIP Conf. Proc., 34 (1976) 307-9 (Mag).

512. H.R.KHAN and Ch.J.RAUB, "Superconductivity of Some Splat-Cooled Nb-Au-Pt Alloys", Z. Physik, B23 (1976) 149-51 (S).

513. L.KRAUS, J.SCHNEIDER and H.WIESNER, "Ferromagnetic Resonance in Amorphous Alloys Prepared by Rapid Quenching from the Melt", Czech. J. Phys., B26 (1976) 601-2 (Mag)(see Chem. Abstr. 85(1976) 170890n).

514. W.D.KRISTENSEN, "Computer-Simulated Amorphous Structures I. Quenching of a Lennard - Jones Model System", J. Non-Cryst. Solids, 21 (1976) 303-18 (SMA).

515. D.KUNSTELJ and K.KRANJC, "Distortion of the Matrix Caused by Coherent Precipitates in the Al-Ni System", Fizika, 8 (1976) 196-8 (AB)(see Phys. Abstr., 80(1977) 59791).

516. D.KUNSTELJ, D.PIVAC, D.ROČAK, M.STUBIČAR and A.BONEFAČIĆ, "Changes of Structure and Microhardness During the Annealing of an Al-Sn Supersaturated Solid Solution", Phil. Mag., 34 (1976) 67-77 (M, AB).

517. V.M.KUZMENKO, V.I.MELNIKOV and A.I.SUDOVTSOV, "Galvanomagnetic Properties and Magnetization of Amorphous Iron Films", Fiz. Nizk. Temp., 2 (1976) 1476-9 (in Russian); Sov. J. Low Temp. phys., 2 (1976) (VD; Mag)(see Phys. Abstr., 80(1977) 35922).

518. M.LARIDJANI, K.D.KRISHNANAND and R.W.CAHN, "An X-Ray Study by Guinier Diffraction Camera of Al-Ge Alloys Rapidly Quenched from the Melt", J. Mater. Sci., 11 (1976) 1643-52 (MCS).

519. M.LASOCKA, "Normalised Value of K_{gl} Parameter of Glass Formation Ability", J. Mater. Sci., 11 (1976) 1770-1 (AS).

520. M.LASOCKA and H.MATYJA, "Physico-Chemical Aspects of the Structure of Non-Crystalline Materials", Arch. Hutn., (Poland) 21 (1976) 449-60 (PM).

521. H.J.LEAMY, E.M.GYORGY, R.C.SHERWOOD, T.WAKIYAMA and H.S.CHEN, "The Effect of Heat Treatment on the Curie Temperature of a Metallic Glass", AIP Conf. Proc., 29 (1976) 211-3 (Mag).

522. J.C.M.LI, "Defect Mechanisms in the Deformation of Amorphous Materials", in 'Frontiers in Materials Science' (Marcel-Dekker, 1976) pp. 527-48 (M, R).

523. H.H.LIEBERMANN and C.D.GRAHAM, Jr., "Production of Amorphous Alloy Ribbons and Effects of Apparatus Parameters on Ribbon Dimensions", IEEE Trans. Mag., MAG-12 (1976) 921-3 (QT).

524. C.C.LO and H.G.TOMPKINS, "Oxidation of Amorphous $Pd_{78}Au_4Si_{18}$ Alloy Studied with Auger Electron Spectroscopy", J. Appl. Phys., 47 (1975) 3496-3502 (C).

525. J.LOGAN and E.SUN, "A Mössbauer Study of Amorphous Iron-Phosphorus Alloys", J. Non-Cryst. Solids, 20 (1976) 285-98 (ED; AQS).

526. J.LOGAN and M.YUNG, "The Low-Temperature Electrical Resistivity of Amorphous Iron-Phosphorus Alloys", J. Non-Cryst. Solids, 21 (1976) 151-4 (ED; E).

527. F.E.LUBORSKY, "Kinetics of Reorientation of Magnetically Induced Anisotropy in Amorphous $Ni_{40}Fe_{40}P_{14}B_6$", AIP Conf. Proc., 29 (1976) 209-10 (Mag).

528. F.E.LUBORSKY and J.L.WALTER, "Stability of Amorphous Metallic Alloys", J. Appl. Phys., 47 (1976) 3648-50 (AS).

529. F.E.LUBORSKY, J.L.WALTER and D.G.LeGRAND, "Cold-rolling and Annealing of Amorphous Ribbons", IEEE Trans. Mag., MAG-12 (1976) 930-2 (Mag).

530. L.H.LUND, "Note on Amorphous Magnetic Materials", J. Chem. Phys., 65 (1976) 1596-8 (Mag).

531. J.W.LYNN, G.SHIRANE, R.J.BIRGENEAU and H.S.CHEN, "Neutron Scattering Study of Spin Waves in the Amorphous Ferromagnet $(Fe_{0.3}Ni_{0.7})_{0.75}P_{0.16}B_{0.06}Al_{0.03}$", AIP Conf. Proc., 34 (1976) 313-5 (Mag).

532. K.MAKI, "Amorphous-Crystalline Transition in Vacuum Deposited Thin Films", J. Vac. Soc. Japan, 19 (1976) 145-54 (in Japanese) (VD; R)(see Phys. Abstr., 79(1976) 83682).

533. P.MANGIN, G.MARCHAL, M.PIECUCH and C.JANOT, "Mössbauer Spectra Analysis in Amorphous System Studies", J. Phys. E: Sci. Instrum., 9 (1976) 1101-5.

534. G.MARCHAL, P.MANGIN and C.JANOT, "Magnetism in Amorphous Fe-Si Alloys", Solid State Commun., 18 (1976) 739-42 (Mag).

535. T.MASUMOTO and T.MURATA, "Deformation of Amorphous Metals", Mater. Sci. & Engg., 25 (1976) 71-5 (R).

536. J.R.MATEY and A.C.ANDERSON, "Thermal Conductivity of a Glassy Metal at Low Temperatures", in Proc. Internat. Conf. Thermal Conductivity, 14 (1976) 31-8 (PP).

537. J.R.McCOLL, D.MURPHY, G.S.CARGILL III and T.MIZOGUCHI, "Spinwave Dispersion and Temperature Dependence of Magnetization in an Amorphous Co-P Alloy", AIP Conf. Proc., 29 (1976) 172-4 (ED; Mag).

538. J.MEGUSAR, J.B.VANDER SANDE and N.J.GRANT, "Mechanical Behavior of Amorphous and Partially Crystalline Alloys", in Proc. Second Internat. Conf. Mechanical Behavior of Materials (ASM, Metals Park, Ohio, USA, 1976) pp. 1164-8 (M)(see Chem. Abstr., 88(1978) 195864g).

539. L.I.MENDELSOHN, E.A.NESBITT and G.R.BRETTS, "Glassy Metal Fabric: A Unique Magnetic Shield", IEEE Trans. Mag., MAG-12 (1976) 924-6 (Mag).

540. M.D.MERZ, S.D.DAHLGREN, R.BUSCH and R.WANG, "Mechanical Properties of Ultrafine-Grained Copper and Copper-Zirconium Sputter Deposits", in Proc. Second Internat. Conf. Mechanical Behavior of Materials (ASM, Metals Park, Ohio, USA, 1976) pp. 58-62 (M).

541. O.MILAT, D.DUŽEVIĆ and A.BONEFAČIĆ, "Levitation Splat-Quenching Device", Fizika. Suppl, 8 (1976) 25-7 (QT)(see Phys. Abstr., 80(1977) 52261).

542. T.MIZOGUCHI, "Magnetism in Amorphous Alloys", AIP Conf. Proc., 34 (1976) 286-91 (Mag, R).

543. T.MIZOGUCHI and T.KUDO, "The Magnetic Properties of 3d Impurities in Amorphous Cu-Zr Alloy", AIP Conf. Proc., 29 (1976) 167-8 (Mag).

544. T.MURATA, H.M.KIMURA and T.MASUMOTO, "Anelastic Strain Recovery of Amorphous Metals", Scripta Met., 10 (1976) 705-9 (M).

545. T.MURATA, H.M.KIMURA and T.MASUMOTO, "Viscoelastic Behavior of Amorphous Metals", in Proc. Second Internat. Conf. Mechanical Behavior of Materials, (ASM, Metals Park, Ohio, USA, 1976) pp. 33-7 (M)(see Chem. Abstr., 88(1978) 53919e).

546. T.MURATA, S.TOMIZAWA, T.FUKASE and T.MASUMOTO, "Electric Properties of Amorphous Cu-Zr Alloy", Scripta Met., 10 (1976) 181-4 (E).

547. S.R.NAGEL, G.B.FISHER, J.TAUC and B.G.BAGLEY, "Photoemission Studies of Crystalline Palladium and the Metallic Alloy $Pd_{.775}$-$Cu_{.06}Si_{.165}$ in its Glassy and Crystalline Forms", Phys. Rev., B13 (1976) 3284-9 (AQS).

548. M.NAKA, K.HASHIMOTO and T.MASUMOTO, "High Corrosion Resistance of Cr-Bearing Amorphous Iron Alloys in Neutral and Acidic Solutions Containing Chloride", Corrosion, 32 (1976) 146-52 (C).

549. G.H.NARAYANAN and R.WANG, "An Electron Microscopy Study of the Metastable Structures in Splat-Cooled Erbrium-Zirconium Alloys", J. Mater. Sci., 11 (1976) 1667-79 (AQS, TSSE, AB).

550. R.OBERLE, H.U.KÜNZI and H.-J.GÜNTHERODT, "Electrical Resistivity of Ternary Glassy and Liquid Pd Alloys", Phys. Letters, A58 (1976) 272-4 (E).

551. Y.OBI, H.FUJIMORI and H.SAITO, "Magnetic Domain Structure of an Amorphous Fe-P-C Alloy", Japan. J. Appl. Phys., 15 (1976) 611-7 (Mag).

552. M.OČKO and E.BABIĆ, "Changes in the Lattice Parameter in Al-3d Alloys", Fizika, 8 (1976) 204 (AQS).

553. M.OČKO, E.BABIĆ, R.KRSNIK, E.GIRT and B.LEONTIĆ, "Some Properties of AlSc Solid Solutions", J. Phys. F: Metal Phys., 6 (1976) 703-7 (TSSE).

554. M.OČKO, E.BABIĆ, and V.ZLATIĆ, "Changes of the Lattice Parameter in Al3d Alloys due to the Virtual Bound State", Solid State Commun., 18 (1976) 705-8 (AQS).

555. R.C.O'HANDLEY, "Magnetic Core Loss and Internal Stress in Metallic Glasses", AIP Conf. Proc., 29 (1976) 206-8 (Mag).

556. R.C.O'HANDLEY and D.S.BOUDREAUX, "Modeling Slater-Pauling Curves", AIP Conf. Proc., 29 (1976) 161 only (SMA).

557. R.C.O'HANDLEY, R.HASEGAWA, R.RAY and C.-P.CHOU, "Ferromagnetic Properties of Some New Metallic Glasses", Appl. Phys. Letters, 29 (1976) 330-2 (Mag).

558. R.C.O'HANDLEY, L.I.MENDELSOHN, R.HASEGAWA, R.RAY and S.KAVESH, "Low-Field Magnetic Properties of $Fe_{80}B_{20}$ Glass", J. Appl. Phys., 47 (1976) 4660-2 (Mag).

559. R.C.O'HANDLEY, L.I.MENDELSOHN and E.A.NESBITT, "New Non-Magnetostrictive Metallic Glass", IEEE Trans. Mag., MAG-12 (1976) 942-4 (Mag).

560. T.OMI, S.KOKUNAI and H.YAMAMOTO, "Structure of Amorphous Ni-P Electrodeposits", Trans. Japan Inst. Metals, 17 (1976) 370-7 (ED; AQS).

561. K.J.OVERSHOTT, "New Magnetic Materials", Electron. & Power, 22 (1976) 768-72 (R)(see Phys. Abstr., 80(1977) 8331).

562. A.F.POLESYA, V.V.KOVALENKO and S.B.VUKELICH, "Crystallization of Rapidly Cooled Al-Ge-Mn Melts", Fiz. Met. Metalloved., 41 (1976) 218-20 (in Russian); Phys. Met. Metallogr., 41, No. 1 (1976) 195-8 (AB).

563. B.PREDEL, "Thermodynamic Properties and Structure of Liquid Alloys", Berichte Bunsen-Gessel., 80 (1976) 695-704 (T).

564. B.PREDEL and H.BANKSTAHL, "Metastable Phases in the Silver-Bismuth Systems", Z. Metallkde., 67 (1976) 793-9 (MCS).

565. I.B.PUCHALSKA and J.F.SADOC, "Zig-Zag Band and Bubble-Like Domains in Co-P Amorphous Thick Films", J. Appl. Phys., 47 (1976) 333-8 (ED; Mag).

566. K.RAJ, J.I.BUNDNICK, R.ALBEN, G.C.CHI and G.S.CARGILL III, "Hyperfine Field Studies of Amorphous Co-P Alloys", AIP Conf. Proc., 31 (1976) 390-6 (ED; Mag).

567. K.V.RAO, R.MALMHÄLL, G.BÄCKSTRÖM and S.M.BHAGAT, "Hall Effect in Amorphous Magnetic Materials- Metglas 2826 and 2826B", Solid State Commun., 19 (1976) 193-5 (Mag).

568. A.REVCOLEVSCHI, "Cooling Rate Determination in Splat Cooling of Oxides", J. Mater. Sci., 11 (1976) 563-5 (PM).

569. L.RIVOALEN, A.REVCOLEVSCHI, J.LIVAGE and R.COLLONGUES, "Amorphous Vanadium Pentoxide", J. Non-Cryst. Solids, 21 (1976) 171-9 (AQS, E).

570. E.M.SAVITSKY, Yu.V.JEFIMOV and G.G.MUKHIN, "Rapid Quenching of Laves Phases of Vanadium with Zr and Hf", Fiz. Met. Metalloved., 41 (1976) 1299-1302 (in Russian); Phys. Met. Metallogr., 41 No. 6 (1976) 152-5 (AQS).

571. A.S.SCHAAFSMA and F.VAN DER WOUDE, "Transformation of Some Amorphous FePC Alloys during Isothermal Ageing", J. Physique, 37 (1976) C6-769 only (AB).

572. J.SCHNEIDER and J.KONETZKY, "Magnetization Studies of Rapidlly Quenched Fe-Ni Alloys", Phys. Stat. Sol., (a) 36 (1976) K77-80 (Mag).

573. J.SCHNEIDER, H.WIESNER and R.GEMPERLE, "Annealing Effects on the Magnetic Properties of Rapidly Quenched Transition Metal Alloys", Phys. Stat. Sol., (a) 36 (1976) K59-64 (Mag).

574. R.C.SHERWOOD, E.M.GYORGY, H.J.LEAMY and H.S.CHEN, "Exchange Anisotropy in Metallic Glasses", AIP Conf. Proc., 34 (1976) 325-7 (Mag).

575. J.W.SHILLING, "Domain Structure during Magnetization of an Annealed and Elastically Strained Amorphous Ni-Fe-P-B Alloy", AIP Conf. Proc., 29 (1976) 222-3 (Mag).

576. Y.SHIMADA and H.KOJIMA, "Magnetic Properties of Amorphous Fe-Si Thin Films", J. Appl. Phys., 47 (1976) 4156-9 (Sp; Mag).

577. A.K.SINHA, B.C.GIESSEN and D.E.POLK, "Metastable Phases Produced by Rapid Quenching from the Vapour and the Liquid" in Treatise on Solid State Chemistry, ed. N.B.Hannay (Plenum Press, New York, USA, 1976) Vol. 3, pp. 1-88 (R).

578. J.SLECHTA, "A Monte-Carlo Study of Binary Magnetic Alloys beyond the Homogeneous Molecular Field Approximation-Anomalous Kondo Effect", J. Non-Cryst. Solids, 22 (1976) 345-58 (SMA).

579. T.SOSHIRODA, M.KOIWA and T.MASUMOTO, "The Internal Friction and Elastic Modulus of Amorphous Pd-Si and Fe-P-C Alloys", J. Non-Cryst. Solids, 22 (1976) 173-87 (M).

580. K.SUZUKI, "Structural Study of Liquids with Strong Short-Range Correlation in the Atomic Distribution", Berichte Bunsen Gessel., 80 (1976) 689-94 (AQS).

581. K.SUZUKI, "Structure of Metals in the Amorphous and Liquid States", Nippon Butsuri Gakkaishi, 31 (1976) 544-7 (in Japanese) (R)(see Chem. Abstr., 85(1976) 147378t).

582. F.R.SZOFRAN, G.R.GRUZALSKI, J.W.WEYMOUTH, D.J.SELLMYER and B.C.GIESSEN, "Electronic and Magnetic Properties of Amorphous and Crystlline $Zr_{40}Cu_{60-x}Fe_x$ Alloys", Phys. Rev., B14 (1976) 2160-70 (E, Mag).

583. L.TAKÁCS, "Mössbauer Investigation of the Magnetic Anisotropy and Electronic Structure of a Metallic Glass", Report KFKI-76-36, Hungarian Acad. Sci., Budapest (June 1976) 7pp. (Mag)(see Phys. Abstr., 79(1976) 72810).

584. M.TAKAHASHI, T.IKEDA and T.MIYAZAKI, "The Variation of the Uniaxial Magnetic Anisotropy along the Thickness in an Amorphous $Fe_{80}P_{13}C_7$", Japan.J. Appl. Phys., 15 (1976) 1821-2 (Mag).

585. M.TAKAHASHI, T.MIYAZAKI, F.ONO and K.TAKAKURA, "The Change of the Magnetic Properties with Annealing for an Amorphous Fe_{80}-$P_{13}C_7$ Alloy", J. Japan Inst. Metals, 40 (1976) 1192-7 (in Japanese) (Mag) (see Phys. Abstr., 80(1977) 16701).

586. M.TAKAHASHI, F.ONO and K.TAKAKURA, "Magnetic Anisotropy of an Amorphous $Fe_{80}P_{13}C_7$ Alloy", Japan.J. Appl. Phys., 15 (1976) 183-4 (Mag).

587. M.TAKAHASHI and T.SUZUKI, "A Study of the Crystallographic Structure for an Amorphous $Fe_{80}P_{13}C_7$ Alloy", Japan. J. Appl. Phys., 15 (1976) 1555-6 (AQS).

588. M.TAKAHASHI, T.SUZUKI and K.TAKAKURA, "The Local Variation of Hysteresis Loops in an Amorphous Fe-P-C Alloy", Japan.J. Appl. Phys., 15 (1976) 711-2 (Mag).

589. Y.TAKAHASHI and M.SHIMIZU, "Magnetic Excitations in Amorphous Ferromagnets", Phys. Letters, A58 (1976) 419-20 (VD; Mag).

590. S.TAKAYAMA, "Amorphous Structures and Their Formation and Stability", J. Mater. Sci., 11 (1976) 164-85 (R).

591. S.TAKAYAMA and R.MADDIN, "Mechanical Behavior of Partially Crystallized Amorphous Alloys", J. Mater. Sci., 11 (1976) 22-6 (M).

592. S.TAKAYAMA and R.MADDIN, "Tensile Deformation of Amorphous Ni-Pd-P Alloys", Met. Trans., A7 (1976) 1065-72 (M).

593. S.TAKAYAMA and R.MADDIN, "Crystallization of Amorphous Ni_{55}-$Pd_{35}P_{10}$ Alloys", J. Non-Cryst. Solids, 20 (1976) 123-9 (AB).

594. J.TAUC and S.R.NAGEL, "Electronic Structure and the Stability of Metallic Glasses", Comments Sol. State Phys., 7 (1976) 69-74 (AS).

595. G.TAYLOR and D.TAYLOR, "A Totally New Kind of Metal (Metallic Glass)", New Scientist, 71 (1976) 323-5 (R)(see Chem. Abstr., 85(1976) 196412h).

596. M.F.THORPE, L.v.HEIMENDAHL and R.ALBEN, "Light Scattering in Metallic Glasses", in Proc. Third Internat. Conf. on Light Scattering in Solids, eds. M.Balkanski, R.C.C.Leite and S.P.S. Porto (Wiley, New York, 1976), pp. 668-72 (PP)(see Chem. Abstr., 85(1976) 181353r).

597. V.I.TKACH, N.I.VARICH and A.A.YAKUNIN, "Structure of Co-Nb Alloys Quenched from the Molten State", Izv. VUZ Tsvetn. Met., No. 1 (1976) 121-5 (in Russian) (AQS)(see Met. Abstr., 9(1976) 12-1350).

598. A.M.TONEJC, A.KIRIN and A.BONEFAČIĆ, "X-Ray Diffraction Study of Splat-Quenched Ag-6 at.%Sn and Ag-11 at.% Sn Alloy", Fizika, 8 (1976) 87-97 (AQS) (see Phys. Abstr., 80(1977) 8860).

599. A.M.TONEJC and D.KUNSTELJ, "Investigation of Vacancy Loops and Precipitates in Al-3d Alloys Rapidly Quenched from the Melt", Fizika, 8 (1976) 194-5 (AQS)(see Phys. Abstr., 80(1977) 59769).

600. C.C.TSUEI and H.LILIENTHAL, "Magnetization Distribution in an Amorphous Ferromagnet", Phys. Rev., B13 (1976) 4899-906 (Mag).

601. N.TSUYA, M.YAMADA, K.I.ARAI, T.MASUMOTO and H.FUJIMORI, "High Frequency Core Loss and Initial Permeability of Ferromagnetic Amorphous Ribbons", Japan. J. Appl. Phys., 15 (1976) 743-4 (Mag).

602. D.TURNBULL, "Formation and Transformation Behavior of Amorphous Solids", NATO Adv. Study Inst. Series, B20 (1976) 1-5 (AS, AB).

603. M.VIARD, R.CHAROTTE and J.FLECHON, "Resistivity of Thin Films of Nickel-Phophorus Deposited by Chemical Methods", Thin Solid Films, 35 (1976) 237-46 (CD; E).

604. V.VISHAN and A.V.NARLIKAR, "Structure and Superconductivity of Splat-Quenched Pb-Bi Solid Solutions", Mater. Res. Bull., 11 (1976) 1257-64 (AQS, S).

605. D.A.VAN CLEAVE, "Rapid Quenching Boosts Metallic Properties", Iron Age, 218 (1976) 52-5.

606. R.K.VISWANADHAM, J.A.S.GREEN and W.G.MONTAGUE, "Hydrogen Embrittlement of an Fe-Based Amorphous Metal", Scripta Met., 10 (1976) 229-30 (M).

607. J.M.VITEK, "Comment on 'The Production of Electron Transparent Areas by Splat Quenching", J. Mater. Sci., 11 (1976) 2153-4 (AQS) (see entry 615).

608. Z.VUČIĆ, B.ETLINGER and D.KUNSTELJ, "Preparation of Pure Glassy Germanium", J. Non-Cryst. Solids, 20 (1976) 451-4 (At; AS).

609. J.L.WALTER, F.BACON and F.E.LUBORSKY, "An Auger Analysis of the Embrittlement of the Amorphous Alloy $Ni_{40}Fe_{40}P_{14}B_6$", Mater. Sci. & Engg., 24 (1976) 239-45 (M).

610. R.WANG and M.D.MERZ, "Non-Crystallinity and Polymorphism in Elemental Solids", Nature, 260 (1976) 35-6 (AS).

611. Y.WASEDA and S.TAMAKI, "Partial Structure Factors of Amorphous Ni-P Alloys", Z. Physik, B23 (1976) 315-9 (AQS).

612. D.WEAIRE, "The Structure of Amorphous Solids", Contemp. Phys., 17 (1976) 173-97 (R).

613. D.B.WILLIAMS and J.W.EDINGTON, "Microanalysis of Splat-Quenched Al-Cu Alloys", Phil. Mag., 34 (1976) 235-42 (AQS).

614. D.B.WILLIAMS and J.W.EDINGTON, "On the Presence of Amorphous Regions in Splat-Quenched Al-Cu Alloys", J. Mater. Sci., 11 (1976) 2146-8 (AQS).

615. D.B.WILLIAMS and J.W.EDINGTON, "The Production of Electron Transparent Areas by Splat Quenching", J. Mater. Sci., 11 (1976) 2151-3 (AQS).

616. R.S.WILLIAMS and T.EGAMI, "Effects of Deformation and Annealing on Magnetic Amorphous Alloys", IEEE Trans. Mag., MAG-12 (1976) 927-9 (Mag).

617. J.V.WOOD and K.N.AKHURST, "The Effect of Gaseous Impurities on the Structure of Splat-Quenched Nickel", J. Mater. Sci., 11 (1976) 2142-5 (PM).

618. J.V.WOOD and V.JACOMBS, "Rapid Quenching of a Commercial Dental Alloy", J. Mater. Sci., 11 (1976) 865-71 (AQS).

619. J.G.WRIGHT, "Amorphous Transition Metal Films", IEEE Trans. Mag., MAG-12 (1976) 95-102 (VD; R).

620. K.YAMAUCHI, "Preparation and Appraisal of Amorphous Magnetic Materials", Oyo Buturi, 45 (1976) 999-1002 (in Japanese) (R) (see Phys. Abstr., 80(1977) 36420).

621. G.ZANICCHI, R.FERRO, R.MARAZZA and V.CONTARDI, "Some Metastable Ag-Rich Alloys in the Ag-Bi-Cu System", J. Less-Common Metals, 50 (1976) 151-4 (MCS).

RAPIDLY QUENCHED METALS (Section I of the Proceedings of the Second International Conference on Rapidly Quenched Metals, eds. N.J.Grant and B.C.Giessen, M.I.T.Press, Cambridge, Mass., U.S.A., 1976)

622. R.P.ALLEN, S.D.DAHLGREN and M.D.MERZ, "Preparation and Characterization of Thick Metastable Sputter Deposits", pp. 37-44 (Sp; AQS).

623. K.I.ARAI, N.TSUYA, M.YAMADA, H.SHIRAE, H.FUJIMORI, H.SAITO and T.MASUMOTO, "Saturation Magnetostriction and Volume Magnetostriction of Amorphous Ribbons Based on Fe-Ni and Fe-Co", pp. 489-94 (Mag).

624. E.BABIĆ, J.R.COOPER, B.LEONTIĆ and M.OČKO, "Interactions and Isolated Impurities in Supersaturated ZnFe Alloys", pp. 495-501 (AQS).

625. E.BABIĆ, R.KRSNIK, B.LEONTIĆ and M.OČKO, "Low Temperature Deviations from Matthiessen's Rule in Rapidly Quenched Aluminum Alloys", pp. 503-9 (E).

626. M.BARMATZ, K.W.WYATT and H.S.CHEN, "Effects of Cold Rolling on Low Temperature Anelastic Properties of Metallic Glasses", pp. 431-9 (M).

627. G.E.A.BARTSCH, P.GLOZBACH and T.JUST, "Measurements of the Short Range Order of Glassy Metallic Alloys Quenched from the Melt by the X-Ray and the Mössbauer Technique and Comparison with the Atomic Distribution", pp. 343-50 (AQS).

628. C.H.BENNETT and J.L.FINNEY, "Structural Changes at the Lennard-Jones Glass Transition", pp. 231-6 (SMA).

629. S.K.BOSE and R.KUMAR, "Rapid Solidification of Al-Mg-Si Alloys from the Liquid State", pp. 169-77 (AQS).

630. D.S.BOUDREAUX and J.M.GREGOR, "Close-Packed Hard Sphere Models of Metallic Glass Alloys", pp. 323-7 (SMA).

631. M.C.CADEVILLE, G.KAUL, M.F.LAPIERRE, C.LERNER and Ph.MAITREPIERRE, "On Metastable Dilute Binary NiB_x and FeC_x Systems Obtained by Splat Quenching", pp. 143-55 (AQS).

632. B.CANTOR and R.W.CAHN, "Metastable Alloy Phases by Getter-Sputtering", pp. 59-65 (Sp; AQS).

633. G.S.CARGILL III, "Current Views on the Structure of Amorphous Metals", pp. 239-321 (SMA).

634. K.CHATTOPADHYAY, P.RAMACHANDRARAO, S.LELE and T.R.ANANTHARAMAN, "Crystal Structure of a Metastable Aluminium-Nickel Phase Obtained by Splat Cooling", pp. 157-61 (MCS).

635. H.S.CHEN and C.C.LO, "Structure Relaxation and Hardness of Metallic Glasses", pp. 413-24 (M).

636. W.G.CLEMENTS and B.CANTOR, "Crystallization of Amorphous Alloys Prepared by Electroless Deposition", pp. 267-72 (CD; AB).

637. H.A.DAVIES and B.G.LEWIS, "The Kinetics of Formation of Metallic Glasses and Its Practical Implications", pp. 259-66 (AS, PM).

638. L.A.DAVIS, "Mechanics of Metallic Glasses", pp. 369-91 (M).

639. G.FANINGER, D.MERZ and H.WINTER, "Investigation of the Corrosion Behavior of Splat-Cooled Aluminium-Iron Alloys with Small Additions of Magnesium", pp. 483-8 (C).

640. M.FISCHER, H.-J.GÜNTHERODT, E.HAUSER, H.U.KÜNZI, M.LIARD and R.MÜLLER, "Electronic Structure of Metallic Glasses", pp. 329-35 (AQS).

641. E.HAUSER, H.U.KÜNZI, M.LIARD and R.MÜLLER, "Electronic Structure of Metallic Glasses", pp. 329-35 (AQS).

642. A.FONTAINE, "Clustering Effects in Splat-Cooled Al-Transition Metal Alloys", pp. 163-7 (AQS).

643. B.C.GIESSEN, "Classification and Crystal Chemistry of Ordered Metastable Alloys Phases", pp. 119-34 (MCS).

644. B.C.GIESSEN, A.E.ATTARD and R.RAY, "Mössbauer Spectroscopy of Amorphous and Crystalline $Zr_xCu_{.925-x}Fe_{.075}$ Alloys", pp. 351-8 (AQS).

645. R.HASEGAWA and R.C.O'HANDLEY, "Magnetization in Some Iron-Base Glassy Alloys", pp. 459-65 (Mag).

646. H.JONES, "Effects of Experimental Variable in Rapid Quenching from the Melt", pp. 1-27 (PM).

647. A.KAWASHIMA, K.HASHIMOTO and T.MASUMOTO, "Stress Corrosion Cracking of Amorphous Iron Base Alloys", pp. 475-81 (C).

648. J.T.KRAUSE and H.S.CHEN, "Uniaxial Stress Dependence of Young's Modulus by An Ultrasonic Technique. Data for Magnetic and Non-Magnetic Metallic Glasses", pp. 425-30 (M).

649. K.D.KRISHNANAND and R.W.CAHN, "Properties of Plasma-Sprayed Aluminium-Copper Alloys", pp. 67-75 (PS; M).

650. H.J.LEAMY, S.D.FERRIS, D.C.JOY, R.C.SHERWOOD, E.M.GYORGY and H.S.CHEN, "The Domain Structure of Ferromagnetic Metallic Glasses", pp. 511-8 (Mag).

651. D.LEE and T.M.DEVINE, "Deformation, Fracture and Corrosion Behavior of An Amorphous Alloy", pp. 393-9 (M).

652. F.E.LUBORSKY, R.O.McCARY and J.J.BECKER, "Magnetic Annealing of Amorphous Alloy Toroids", pp. 467-73 (Mag).

653. R.E.MARINGER, C.E.MOBLEY and E.W.COLLINGS, "An Experimental Method for The Casting of Rapidly Quenched Filaments and Fibers", pp. 29-36 (QT).

654. T.B.MASSALSKI and Y.BIENVENU, "Metastable Phase Relationships Produced by Rapid Freezing of Alloys of Zn and Cd with Cu and Ag", pp. 95-118 (TSSE).

655. J.MEGUSAR, J.VANDER SANDE and N.J.GRANT, "Crystallinity and Mechanical Properties of Metallic Glasses", pp. 401-11 (M).

656. S.R.NAGEL and J.TAUC, "Stability of Metallic Glasses: Photoemission Studies and A New Electronic Model for Glass Stability", pp. 337-42 (SMA).

657. M.NAKA, S.TOMIZAWA, T.MASUMOTO and T.WATANABE, "Effects of Alloying Elements on Strength and Thermal Stability of Amorphous Iron-Base Alloys", pp. 273-9 (M, T).

658. D.H.RASMUSSEN, J.H.PEREPEZKO and C.R.LOPER, Jr., "Studies of The Undercooling and Nucleation of Liquid Metals and Alloys Utilizing The Droplet Technique", pp. 51-8 (QT).

659. A.REVCOLEVSCHI and R.COLLONGUES, "Application of The Splat-Cooling Method to Non-Metallic Systems", pp. 77-86 (QT).

660. I.R.SARE and R.W.K.HONEYCOMBE, "Structures Produced by Rapid Quenching of Molybdenum Steels", pp. 179-87 (AQS).

661. M.G.SCOTT and R.MADDIN, "Comparisons between Vapor- and Liquid-Quenched Amorphous Alloys", pp. 249-58 (AQS).

662. P.H.SHINGU, K.SHIMOMURA, K.KOBAYASHI and R.OZAKI, "Solidification Rate of Rapidly Quenched Metals", pp. 45-50 (PM).

663. F.SPAEPEN and D.TURNBULL, "Formation of Metallic Glasses", pp. 205-29 (AS).

664. C.C.TSUEI, "Electrical and Magnetic Properties of Rapidly Quenched Metals", pp. 441-58 (E, Mag).

665. K.TOGANO and K.TACHIKAWA, "Structure and Superconductivity of Metastable Phases in Liquid Quenched Zr-Rh Alloys", pp. 519-26 (S).

666. C.N.J.WAGNER, "The Structure of Amorphous Co-Gd Films Prepared by Bias Sputter Deposition from a Co_3Gd Target", pp. 359-67 (Sp; AQS).

667. R.WANG and S.D.DAHLGREN, "Metastable A15 Phase Superconductors Transformed from Metastable BCC Nb_3X Solid Solutions Made by High-Rate Sputter Deposition", pp. 189-96 (Sp; S).

668. H.WARLIMONT and P.KUNZMANN, "Effects of Rapid Solidification in Arc Sprayed Aluminium Alloys", pp. 197-204 (AQS).

669. D.B.WILLIAMS and J.W.EDINGTON, "High Resolution Microanalysis and Micorstructural Characteristics of Splat Quenched Aluminium-Copper Alloys", pp. 135-42 (AQS).

670. J.V.WOOD and I.R.SARE, "Splat Quenching of Iron-Base Alloys in a Controlled Atmosphere", pp. 87-94 (QT).

671. H.L.YEH and R.MADDIN, "Phase Transformations in Amorphous Fe-P-C Alloys", pp. 281-91 (AB).

672. P.G.ZIELINSKI and H.MATYJA, "Influence of Liquid Structure on Glass Forming Tendency", pp. 237-48 (PM).

Mater. Sci. & Engg., 23 (1976) (Section II of the Proceedings of the Second International Conference on Rapidly Quenched Metals, eds. N.J.Grant and B.C.Giessen)

673. A.S.ARGON and M.SALAMA, "The mechanism of Fracture in Glassy Materials Capable of Some Inelastic Deformation", pp. 219-30 (M).

674. D.G.AST and D.KRENITSKY, "Fracture Toughness and Yield Strength of Annealed Ni-Fe Base Metallic Glasses", pp. 241-6 (M).

675. R.W.CAHN, K.D.KRISHNANAND, M.LARIDJANI, M.GREENHOLZ and R.HILL, "Novel Splat-Quenching Techniques and Methods for Assessing Their Performance", pp. 83-6 (QT).

676. H.S.CHEN, "Glass Temperature, Formation and Stability of Fe, Co, Ni, Pd and Pt Based Glasses", pp. 151-4 (AS, T).

677. G.C.CHI and G.S.CARGILL III, "Annealing Effects in Amorphous Co-P Alloys", pp. 155-9 (ED; AB).

678. E.COLEMAN, "Crystallization of Fe, Co and Ni Based Metallic Glasses", pp. 161-7 (AB).

679. H.A.DAVIES and J.B.HULL, "Amorphous Nickel Produced by Splat Quenching", pp. 193-8 (AS).

680. M.DOI, M.YOSHIDA, M.NONOYAMA, T.IMURA, Y.YASHIRO and T.MASUMOTO, "Structure and Stability of a Splat-Cooled Fe-P-C Alloy", pp. 169-72 (AS, AB).

681. J.P.H.A.DURAND, R.M.PELLOUX and N.J.GRANT, "Properties of Splat-Quenched 7075 Aluminium Type Alloys", pp. 247-56 (M).

682. S.EWERT, A.COMBERT and H.WÜHL, "Specific Heat and Tunneling Measurements on Amorphous Indium and Thallium Films", pp. 275-6 (VD; T).

683. G.FANINGER, D.MERZ and H.WINTER, "Preliminary Experiments for The Development of Ductile Superconducting Alloys on The Base of Copper by Rapid Solidification from The Melt", pp. 277-80 (S).

684. J.L.FINNEY, "Fine Structure in Randomly Packed, Dense Clusters of Hard Spheres", pp. 199-205 (SMA).

685. H.FUJIMORI, Y.OBI, T.MASUMOTO and H.SAITO, "Soft Ferromagnetic Properties of Some Amorphous Alloys", pp. 281-4 (Mag).

686. B.C.GIESSEN and R.O.ELLIOTT, "Two New Np-Ga Phases: α-$NpGa_2$ and Metastable m-$NpGa_2$", pp. 113-8 (MCS).

687. B.C.GIESSEN, M.MADHAVA, D.E.POLK and J.B.VANDER SANDE, "Refractroy Amorphous Inter-Transition Metal Alloys", pp. 145-50 (AS).

688. E.GIRT, Z.GATALO, M.IVANOVIĆ, A.KURŠUMOVIĆ and B.LEONTIĆ, "Investigation of Rapidly Quenched Alloy Samples by Simultaneous Resistometric and X-ray Diffraction Measurements", pp. 91-4 (AQS).

689. D.J.GOLDWASSER and B.H.KEAR, "Properties of a Resin Matrix Composite Reinforced with Amorphous Metal Ribbons", pp. 237-9 (M).

690. A.HAMZIĆ, E.BABIĆ and B.LEONTIĆ, "The Impurity Resistivity and Interactions in Supersaturated Al(Mn) and Al(Cr) Alloys", pp. 271-4 (E).

691. R.HASEGAWA, "Further Evidence for the $T^{-1/2}$ Sigularity in Amorphous Kondo Alloys", pp. 293-6 (E).

692. K.HASHIMOTO and T.MASUMOTO, "Ectremely High Corrosion Resistance of Chromium-Containing Amorphous Iron Alloys", pp. 285-8 (C).

693. A.HIRAKI, M.IWAMI, A.SHIMIZU and K.SHUTO, "Metallic State of Si in Rapidly Quenched Si-Noble Metal Alloys", pp. 289-92 (AQS).

694. L.KAUFMAN and H.NESOR, "Application of Computer Techniques to Prediction of Metastable Transitions in Metallic Systems", pp. 119-23 (PM).

695. H.O.K.KIRCHNER, "Microcalorimetry of Metastable Alloys", pp. 95-6 (T).

696. D.C.KOSKENMAKI, "A Model for the Structure of Amorphous Metals", pp. 207-10 (SMA).

697. M.LARIDJANI and R.W.CAHN, "Identification of Metastable Phases in Splat-Quenched Aluminium-Germanium Alloys", pp. 125-9 (MCS).

698. M.LASOCKA, "The Effect of Scanning Rate on Glass Transition Temperature of Splat-Cooled $Te_{85}Ge_{15}$", pp. 173-7 (T).

699. B.G.LEWIS and H.A.DAVIES, "The Stabilities and Kinetics of Formation of Glassy Palladium-Silicon Phases in the Composition Range 5-25 at.% Si", pp. 179-82 (AS, AB).

700. J.LIVAGE and R.COLLONGUES, "Semiconducting Properties of Amorphous V_2O_5 Prepared by Splat Cooling", pp. 297-9 (E).

701. M.MARCUS and D.TURNBULL, "On the Correlation between Glass-Forming Tendency and Liquidus Temperature in Metallic Alloys", pp. 211-4 (PM).

702. T.MASUMOTO, H.KIMURA, A.INOUE and Y.WASEDA, "Structural Stability of Amorphous Metals", pp. 141-4 (AS).

703. H.MATSUEDA and B.L.AVERBACH, "Sputtered Films of Nickel and Nickel Oxide", pp. 131-4 (Sp; AQS).

704. M.NAGUMO and T.TAKAHASHI, "Hydrogen Embrittlement of Some Fe-Base Amorphous Alloys", pp. 257-9 (M).

705. W.B.NOWAK, "Microcrystalline/Amorphous Iron Alloy Films for Corrosion-Resistant Coatings", pp. 301-5 (Sp; A).

706. T.OGURA, K.FUKUSHIMA and T.MASUMOTO, "Propagation of Fatigue Cracks in Amorphous Metals", pp. 231-5 (M).

707. R.P.POND, Sr. and J.M.WINTER, Jr., "Calorimetric Comparison of Quench Rates in Melt Spinning and Melt Extraction", pp. 87-9 (T).

708. J.F.SADOC and J.DIXMIER, "Structural Investigation of Amorphous CoP and NiP Alloys by Combined X-ray and Neutron Scattering", pp. 187-92 (ED; AQS).

709. P.H.SHINGU, K.SHIMOMURA, K.KOBAYASHI and R.OZAKI, "An Early Stage of Crystallization of Amorphous Fe-C Binary Alloys Obtained by Splat Cooling", 183-6 (AB).

710. K.SUZUKI, T.FUKUNAGA, M.MISAWA and T.MASUMOTO, "Time-of-Flight Pulsed Neutron Diffraction of $Pd_{0.8}Si_{0.2}$ Amorphous Alloy Using the Electron Linac", pp. 215-8 (SMA; AQS).

711. W.M.SWIFT and K.FOSTER, "A. C. Losses and Temperature-Dependent Magnetic Properties of an Amorphous Magnetic Alloy", pp. 267-70 (Mag).

712. S.TAKAYAMA and R.MADDIN, "Microscopic Observations of Fracture Behavior in a $Ni_{55}Pd_{35}P_{10}$ Metallic Glass", pp. 261-5 (M).

713. R.WANG, "Solubility and Stability of Liquid-Quenched Metastable h.c.p. Solid Solutions", pp. 135-40 (TSSE).

714. H.WARLIMONT, W.ZINGG and P.FURRER, "Structure of Splat-Cooled Al-Cr Alloys", pp. 101-5 (AQS).

715. T.WATANABE and Y.TANABE, "Formation and Morphology of Ni-B Amorphous Alloy Deposited by Electroless Plating", pp. 97-100 (CD; AS, AQS).

716. J.M.WILLIAMS, J.B.DUNLOP and B.LEONTIĆ, "^{57}Fe Mössbauer Measurements in Zn(Fe) Alloys", pp. 307-8 (Mag).

717. J.V.WOOD and R.W.K.HONEYCOMBE, "Splat Quenching of Austenitic Steels", pp. 107-12 (AQS).

Special Issue on 'CHARACTERISTICS OF AMORPHOUS METALS' Bull. Japan Inst. Metals, 15 (1976) (All articles in Japanese).

718. H.FUJIMORI, "On the Properties of Amorphous Metals as Soft Magnetic Materials", pp. 188-94 (Mag).

719. F.E.FUJITA, "On the Crystalline Nature of Amorphous Metals", pp. 180-3 (AQS).

720. K.HASHIMOTO, "Corrosion Resistance of Amorphous Alloys", pp. 203-6 (C).

721. T.ICHIKAWA, "Structure of Amorphous Metallic Solids", pp. 159-62 (SMA).

722. N.S.KAZAMA, "Some Interesting Magnetic Properties of Amorphous Metals", pp. 184-7 (Mag).

723. T.MASUMOTO, "Characteristics of Amorphous Metals", pp. 151-4(R).

724. T.MASUMOTO, "Mechanical Properties and Deformation-Fracture of Amorphous Metals", pp. 194-202 (M).

725. T.MASUMOTO and Y.WASEDA, "Structural Stability and Crystallization of Amorphous Metals", pp. 175-9 (AQS, AB).

726. P.H.SHINGU and R.OZAKI, "Formation of Amorphous Metals", pp. 171-5 (AS, PM).

727. K.SUZUKI, "Structural Characteristics Between Metallic Glasses and Liquids", pp. 162-7 (SMA).

728. M.TANAKA, "Theoretical Aspects of Structure of Amorphous Metals", pp. 155-8 (AQS).

729. Y.WASEDA, "Current Views on the Structure of Amorphous Metals", pp. 167-71 (SMA, AQS).

AMORPHOUS MATERIALS ISSUE I (Sci. Rep. RITU, A26(1976))

730. H.FUJIMORI, M.KIKUCHI, Y.OBI and T.MASUMOTO, "New Co-Fe Amorphous Alloys as Soft Magnetic Materials", pp. 36-47 (Mag).

731. K.HASHIMOTO, M.NAKA and T.MASUMOTO, "Effect of Nickel Addition on Corrosion Resistance of Amorphous Iron-Base Alloys", pp. 48-54 (C).

732. T.MASUMOTO, Y.WASEDA, H.M.KIMURA and A.INOUE, "Thermal Instability and Crystallization Characteristics of Amorphous Metal-Metalloid System", pp. 21-35 (AB, T, M).

733. K.SUZUKI, T.FUKUNAGA, M.MISAWA and T.MASUMOTO, "Short-Range Structures of Amorphous and Liquid Iron and $Pd_{0.8}Si_{0.2}$ Alloy", pp. 1-11 (AQS).

734. Y.WASEDA, H.OKAZAKI, M.NAKA and T.MASUMOTO, "Structural Study of an Amorphous Liquid-Quenched $Ni_{80}P_{20}$ Alloy", pp. 12-20 (AQS).

1977

735. C.N.AFONSO, F.BRIONES and J.L.VINCENT, "Even Magneto-Optic Effects in a Ferromagnetic Amorphous Alloy", Physica, 89 B+C (1977) 271-2 (Mag).

736. S.C.AGARWAL and H.HERMAN, "Liquid-Quenched Ag-Ge Alloys; Formation and Evolution of Metastable Phases", J. Mater. Sci., 12 (1977) 2021-7 (MCS).

737. R.A.AKOPYAN and A.A.ABDULLAEV, "Measuring the Velocity of Ultrasonic Waves as a Means of Analyzing Breakup of Solid Solutions", Ind. Lab., 43 (1977) 224-7 (AB)(see Met. Abstr., 11(1978) 22-0200).

738. R.A.AKOPYAN, A.V.YEVDOKMOV and L.M.PAVLOVA, "Metstable Continuous Solid Solutions in the Ge-GaSb System", Izv. Akad. Nauk SSSR Mater., 13 (1977) 1970-3 (in Russian)(see Met. Abstr., 11 (1978) 11-0230 (TSSE).

739. R.A.AKOPYAN, A.V.YEVDOKMOV and B.N.USHKOV, "The Breakdown Kinetics of the Metastable Solid Solutions $Ge_{1-x}(GaSb)_x$ Obtained by the Method of Diffusionless Crystallization", Zhur. Fiz. Khim., 51 (1977) 3113-6 (in Russian)(AB)(see Met. Abstr., 11(1978) 15-1295).

740. A.ALI, W.A.GRANT and P.J.GRUNDY, "Amorphous Ni-P Alloys Prepared by Ion Implantation", Radiation Effects, 34 (1977) 251-4 (II; AS).

741. P.ALLIA, F.FIORILLO, G.P.SOARDO and F.VINAI, "Amorphous Magnetic Materials", in Rend. Ruin Ann. Assoc. Elettrotech. Elettron. Ital, 1977, Paper No. 8, 7pp (in Italian)(see Chem. Abstr., 90(1979)161238g).

742. P.ALLIA, G.P.SOARDO and F.VINAI, "Torque Magnetometer Measurements of the Temperature Dependence of Induced Anisotropy Energy and of Saturation Magnetization in Amorphous $Fe_{40}Ni_{40}P_{14}B_6$", Solid State Commun., 24 (1977) 517-9 (Mag).

743. T.R.ANANTHARAMAN and K.CHATTOPADHYAY, "Structure and Properties of Rapidly Solidified Steels and Ferrous Alloys", Tool & Alloy Steels, 11 (1977) 19-27 (R).

744. T.R.ANANTHARAMAN, P.RAMACHANDRARAO, C.SURYANARAYANA, S.LELE and K.CHATTOPADHYAY, "Structure and Constitution of Rapidly Solidified Aluminium Alloys-I. General Assessment", Trans. Indian Inst. Metals, 30 (1977) 423-33 (R).

745. T.R.ANANTHARAMAN, P.RAMACHANDRARAO, C.SURYANARAYANA, S.LELE and K.CHATTOPADHYAY, "Structure and Constitution of Rapidly Solidified Aluminium Alloys-II Annotated Bibliography", Trans. Indian Inst. Metals, 30 (1977) 434-48 (R).

746. P.M.ANDERSON III and A.E.LORD,Jr, "Effect of Hydrostataic Pressure on the Magnetic Properties of Certain Metallic Glasses", J. Appl. Phys., 48 (1977) 4839-43 (Mag).

747. K.I.ARAI and N.TSUYA, "Change of Permeability in Amorphous Ferromagnetic Ribbons by Heat Treatment", IEEE Trans. Mag., MAG-13 (1977) 1550-2 (Mag, PP).

748. J.D.AXE, G.SHIRANE, T.MIZOGUCHI and K.YAMAUCHI, "Neutron Scattering Study of Spin Waves in the Amorphous Metallic Ferromagnet $(Fe_{93}Mo_7)_{80}B_{10}P_{10}$", Phys. Rev., B15 (1977) 2763-70 (Mag).

749. E.BABIĆ, "New Soft Amorphous Ferromagnetic Fe-Ni Alloys", Electrotehnica, Zagreb, No. 5-6 (1977) 377-86 (in Croatian) (Mag)(see Phys. Abstr., 82 (1979) 11298).

750. T.W.BARBEE, Jr., W.H.HOLMES, D.L.KEITH, M.K.PYZYNA and Gh. ILONCA, "Synthesis of Amorphous Nb-Ni Alloys by Vapour Quenching", Thin Solid Films, 45 (1977) 591-9 (VD; AS).

751. J.J.BECKER, F.E.LUBORSKY and J.L.WALTER, "Magnetic Moments and Curie Temperatures of $(Fe, Ni)_{80}(P,B)_{20}$ Amorphous Alloys", IEEE Trans. Mag., MAG-13 (1977) 988-91 (Mag).

752. G.BELLESSA, "Logarithmic Dependence of the Sound Velocity in an Amorphous Metal", Rev. Physique Appl., 12 (1977) 817 only. (in French) (ED; PP)(see Phys. Abstr., 80(1977) 72659).

753. G.BELLESSA, "Sound Propagation in Amorphous Co-P", J. Phys. C: Solid State Phys., 10 (1977) L285-7 (ED; PP).

754. G.BELLESSA and O.BETHOUX, "Sound Propagation in Amorphous Metallic Pd-Si at Low Temperatures", Phys. Letters, A62 (1977) 125-6 (ED; PP).

755. G.BELLESSA, P.DOUSSINEAU and A.LEVELUT, "Logarithmic Temperature Dependence of the Sound Velocity in Amorphous Metallic NiP", J. Physique Letters, 38 (1977) L65-6 (ED; PP).

756. A.BERRADA, F.GAUTIER, M.F.LAPIERRE, B.LOEGEL, P.PANISSOD, C.ROBERT and J.BEILLE, "Magnetic Properties of Amorphous Ni-P Alloys", Solid State Commun., 21 (1977) 671-4 (ED; Mag).

757. B.S.BERRY and W.C.PRITCHET, R.RAY and B.C.GIESSEN, "Internal Friction of Splat-Cooled Gd-Cu and Gd-Fe Alloys", Solid State Commun., 21 (1977) 305-7 (M).

758. S.M.BHAGAT, S.HARALDSON and O.BECKMAN, "Ferromagnetic Resonance in Amorphous Systems. Dependence of the Relaxation Parameter on Temperature (Magnetization)", J. Phys. Chem. Solids, 38 (1977) 593-9 (Mag).

759. A.BONEFAČIĆ, D.KUNSTELJ and M.STUBIČAR, "Decomposition of Aluminium-Based Al-Ni, Al-Sn and Al-Ag Solid Solutions Quenched from the Melt", Wissen. Zeit. Pedagog. Hochschule, N.K.Krupskaja, 15 (1977) 22 (AB).

760. V.T.BORISOV, A.I.DUKHIN, A.M.DURACHENKO, L.A.ALEKSEEV and V.I.KIRIENKO, "Formation of Amorphous Alloys on Quenching from the Melt", Izv. V.U.Z. Chernaya Metall., No. 3 (1977) 127-31 (AS, PM).

761. P.G.BOSWELL, "The Glass Transition Temperature of a $(Ni_5Pd_5)_{82}$-P_{18} Amorphous Alloy", Scripta Met., 11 (1977) 603-6 (T).

762. P.G.BOSWELL, "Crystallization of a $(Ni_5Pd_5)_{82}P_{18}$ Amorphous Alloy", Scripta Met., 11 (1977) 701-7 (AB).

763. P.G.BOSWELL, "The Significance of the Crystallization Temperature of a Glass", Mater. Res. Bull., 12 (1977) 331-40 (T).

764. P.G.BOSWELL and G.A.CHADWICK, "Eutectoid Decomposition of Zn-Al Splat-Quenched and Bulk Specimens", Acta Met., 25 (1977) 779-92 (AB).

765. P.G.BOSWELL and G.A.CHADWICK, "The Structure of γ' Extended Solid Solution in a Splat-Cooled Ag-50 at.% Cu Alloy", J. Mater. Sci., 12 (1977) 1879-94 (TSSE, AQS).

766. P.G.BOSWELL and G.A.CHADWICK, "The Grain Size of Splat-Quenched Alloys", Scripta Met., 11 (1977) 459-65 (PM).

767. D.S.BOUDREAUX and J.M.GREGOR, "Structure Simulation of Transition Metal-Metalloid Glasses", J. Appl. Phys., 48 (1977) 152-8 (SMA).

768. D.S.BOUDREAUX and J.M.GREGOR, "Structure Simulation of Transition Metal-Metalloid Glasses. III", J. Appl. Phys., 48 (1977) 5057-61 (SMA).

769. H.A.BROOKS and H.S.CHEN, "Magnetostriction as a Function of Glass Formers in Amorphous Alloys of Fe-P-B-Al", J. Appl. Phys., 48 (1977) 4837-8 (Mag).

770. W.E.BROWER and S.N.SINGH, "Difficulty of Coarsening of Silica Inclusions in Slowly and Rapidly Solidified Iron", Met. Trans., A8 (1977) 1010-3 (AB).

771. D.K.BROWN, I.NOWIK and D.I.PAUL, "A Mössbauer Study of Ferromagnetic Domain Orientation in Amorphous $Fe_{75}P_{15}C_{10}$ Foils", Solid State Commun., 24 (1977) 711-5 (Mag).

772. E.BURZO, "On the Magnetic Behavior of RT_x (T=Fe or Co) Amorphous Rare Earth Compounds", Rev. Roum. Phys., 22 (1977) 607-12 (Mag) (see Phys. Abstr., 81(1978) 19221).

773. K.H.J.BUSCHOW, M.BROUHA, J.W.M.BIESTERBOS and A.G.DIRKS, "Crystalline and Amorphous Rare-Earth Transition Metal Alloys", Physica, 91B (1977) 261-70 (Mag).

774. M.C.CADEVILLE, J.M.FRIEDT and C.LERNER, "Structural, Electronic and Magnetic Properties of Splat Quenched FeC_x Alloys ($x \leq 0.05$)", J. Phys. F: Metal Phys., 7 (1977) 123-37 (E, Mag).

775. R.W.CAHN, "Irrepressible Superconductor", Nature, 269 (1977) 198-9 (S).

776. A.CALKA, M.MADHAVA, D.E.POLK, B.C.GIESSEN, H.MATYJA and J.B.VANDER SANDE, "A Transition-Metal-Free Amorphous Alloy : $Mg_{0.70}Zn_{0.30}$", Scripta Met., 11 (1977) 65-70 (AS).

777. E.CALLEN, Y.J.LIU and J.R.CULLEN, "Initial Magnetization, Remanence and Coercivity of the Random Anisotropy Amorphous Ferromagnet", Phys. Rev., B16 (1977) 263-70 (Mag).

778. G.S.CARGILL III, "Magnetism in Amorphous Metals and Alloys", Physica, 91B+C (1977) 177-8 (Mag).

779. W.A.CENSE, M.KLERK and W.ALBERS, "Aligned In-Situ Growth of the Fibrous $Ni-Nb_2O_5$ Composite by Preferential Internal Oxidation of Amorphous Nb_2Ni_3", J. Mater. Sci., 12 (1977) 2184-8 (AQS).

780. B.T.A.CHANG and J.C.M.Li, "Irradiation Swelling of Amorphous $Fe_{40}Ni_{40}P_{14}B_6$ Alloy", Scripta Met., 11 (1977) 933-6.

781. H.CHANG and S.SASTRI, "Crystallization of an Fe-Ni base Amorphous Alloy", Met. Trans., A8 (1977) 1063-70 (AB).

782. H.S.CHEN, "Thermal and Mechanical Stability of Metallic Glass Ferromagnets", Scripta Met., 11 (1977) 367-70 (T, M).

783. H.S.CHEN, "Structure and Properties of Metallic Glasses", in Proc. Internat. Symp. Structure of Non-Crystalline Materials, ed. P.H.Gaskell (Taylor & Francis, London, 1977) pp. 79-84 (R) (see Chem. Abstr., 90 (1979) 42313t).

784. H.S.CHEN and S.Y.CHUANG, "Investigation of Atomic Clustering in Aged Metallic Glasses Using Positron Annihilation Methods", Appl. Phys. Letters, 31 (1977) 255-6 (AB).

785. H.S.CHEN and J.T.KRAUSE, "Correlation between Young's Modulus and Thermal Properties of Metallic Glasses", Scripta Met., 11 (1977) 761-4 (M, T).

786. H.S.CHEN, R.C.SHERWOOD and E.M.GYORGY, "The Influence of Composition and Ageing on the Curie Temperature of Metallic Glasses", IEEE Trans. Mag., MAG-13 (1977) 1538-40 (Mag).

787. M.C.CHI and R.ALBEN, "Hysteresis Curves and Magnetization Processes in a Model for an Amorphous Magnet with Random Uniaxial Anisotropy", J. Appl. Phys., 48 (1977) 2987-91 (Mag).

788. C.L.CHIEN and R.HASEGAWA, "$T^{3/2}$ Dependence of Hyperfine Field and Spin-Wave Excitations in Ferromagnetic Metallic Glasses", Phys. Rev., B15 (1977) 2115-23 (Mag).

789. C.L.CHIEN and R.HASEGAWA, "Mössbauer Study of Glassy Ferromagnet: $Fe_{40}Ni_{40}P_{14}B_6$", Phys. Rev., B16 (1977) 3024-35 (Mag).

790. C.L.CHIEN, D.MUSSER, F.E.LUBORSKY, J.J.BECKER and J.L.WALTER, "Magnetic Hyperfine Interaction and Curie Temperature of Amorphous $(Fe_xNi_{1-x})_{80}P_{14}B_6$", Solid State Commun., 24 (1977) 231-5 (Mag).

791. C.P.CHOU, L.A.DAVIS and M.C.NARASIMHAN, "Elastic Constants of Metallic Glasses", Scripta Met., 11 (1977) 417-23 (M).

792. M.CHYCZEWSKI, B.SZYMAJDA and H.MATYJA, "Metastable Intermediate Phases Formed During Very Fast Cooling of Liquid Zn-Cu Alloys", Metalozn. Obrobka Cieplna, 25 (1977) 2-4 (in Polish) (MCS) (see Chem. Abstr., 87(1977) 121423v).

793. C.F.CLINE and R.W.HOPPER, "Explosive Fabrication of Rapidly Quenched Materials", Scripta Met., 11 (1977) 1137-8 (F).

794. R.W.COCHRANE and J.O.STROM-OLSEN, "Temperature and Magnetic Field Dependence of the Resistivity of Amorphous Alloys", J. Phys. F: Metal Phys., 7 (1977) 1799-1810 (ED; E).

795. M.M.COLLVER and R.H.HAMMOND, "Reduced Superconducting Transition Temperatures in Amorphous Transition Metal Alloys", Solid State Commun., 22 (1977) 55-7 (VD; S).

796. P.J.COTE, G.P.CAPSIMALIS and L.V.MEISEL, "Reversible Effects of Temperature on the Structure of Amorphous NiP", Phys. Rev., B16 (1977) 4651-4 (ED; AQS).

797. P.J.COTE and L.V.MEISEL, "Resistivity in Amorphous and Disordered Crystalline Alloys", Phys. Rev. Letters, 39 (1977) 102-5 (E).

798. R.M.J.COTTERILL and F.KRAGH, "Field-Ion Microscopy of Glassy $Pd_{80}Si_{20}$", Scand. J. Metallurgy, 6 (1977) 191-2 (AQS).

799. H.A.DAVIES, "Recent Researches in Metallurgy-Metallic Glasses", Proc. Annual Conf. Scientific Soc. Wales, eds. I. ap Gwynn and I.O.Morris (1977) 25-49 (R) (in Welsh).

800. E.A.DAVIS, "Non-Crystalline Materials", Endeavour, (1977) 103-6 (R)(see Chem. Abstr., 88(1978) 125221 w).

801. L.A.DAVIS, "Metallic Glasses", in Fundamental Aspects of Structural Alloy Design, eds. by R.I.Jaffee and B.A.Wilcox (Plenum Press, New York, 1977) pp. 431-50 (M).

802. M.DEMIT and K.LÖHBERG, "Crystallization of Metallic Melts under the Effect of a Shock Wave. I.Measurement of the Flight Speed of a Melt Sprayed by a Shock-Wave.", Materialprufung, 19 (1977) 350-1 (in German) (PM)(see Met. Abstr., 11(1978) 12-0427).

803 T.M.DEVINE, "Anodic Polarization and Localized Corrosion Behaviour of Amorphous $Ni_{35}Fe_{30}Cr_{15}P_{14}B_6$ in Near Neutral and Acidic Chloride Solutions", J. Electrochem. Soc., 124 (1977) 38-44 (C).

804. G.DIETZ, "Amorphous Ferromagnetic Materials Deposited from Vapour or Liquids", J. Mag. Mag. Mater., 6 (1977) 47-51 (Mag, R).

805. G.DIETZ and Th.FRECHEN, "Time Effects in Ferromagnetic Amorphous Co-P Alloys", J. Mag. Mag. Mater., 6 (1977) 65-7 (ED; AB).

806. G.DIETZ and A.HÜNSELER, "Domain Structure in Amorphous Co-P Alloys", J. Mag. Mag. Mater., 6 (1977) 68-72 (Mag).

807. G.DIETZ, U.KULLMANN and G.LEDERER , "The Complex Initial Permeability of Amorphous Co-P Foils", J. Mag. Mag. Mater., 4 (1977) 30-24 (ED; Mag).

808. J.DIXMIER, "Ideal Amorphous Structure and Structure 'Defects' in Amorphous Substances", J. Calorim., Anal. Therm., 8, Suppl. (1977) 45-56 (in French) (R)(see Chem. Abstr., 88(1978) 180343n).

809. K.DOI, H.KAYANO and T.MASUMOTO, "Structural Changes in Amorphous $Pd_{80}Si_{20}$ by Neutron Irradiation", Appl. Phys. Letters, 31 (1977) 421-2 (AB).

810. P.DOUSSINEAU, A.LEVELUT, G.BELLESSA and O.BETHOUX,"Ultrasonic Attenuation in the Amorphous Metals NiP and PdSi at Low Temperatures", J. Physique Letters, 38 (1977) L483-7 (ED; PP).

811. J.M.DUBOIS and G.LE CAER, "Electron Diffraction and Mössbauer Studies of the ε Phase Retained in Splat-Quenched Fe-C and Fe-C-Si Alloys", Acta Met., 25 (1977) 609-18 (Mag, AQS).

812. A.M.DURACHENKO, I.D.VETOOSHKIN, L.A.ALEKSEEV, A.I.DUKHIN and V.T.BORISOV, "Amorphous Fe-P-C Films Obtained by Quenching from the Liquid State", Fiz. Magn. Plenok (Irkutsk), No. 10 (1977) 152-7 (in Russian) (Mag)(see Met. Abstr., 12(1979) 33-2285).

813. J.DURAND and S.J.POON, "Magnetic and Transport Properties of Amorphous Ferromagnetic Gd-Au, Gd-Ni and Gd-Co Alloys Obtained by Splat Cooling", IEEE Trans. Mag.,MAG-13 (1977) 1556-8 (Mag).

814. P.DUWEZ, "Glass and Glass Products", in McGraw-Hill Yearbook of Science & Technology (1977) 272-4 (R).

815. T.EGAMI, "Low-Field Magnetic Properties of Amorphous Alloys", J. Amer. Ceramic Soc., 60 (1977) 128-33 (Mag).

816. H.EIFRIG, W.GROSSE-NOBIS and K.JANSEN, "Propagation of the Magnetizing Processes Along Thin Strips of a Metallic Glass", J. Mag. Mag. Mater., 6 (1977) 73-6 (Mag).

817. H.F.ENO, E.H.TYLER and H.L.LUO, "Specific Heats of Fast-Quenched Au-Ni Alloys", J. Low Temp. Phys., 28 (1977) 443-8 (T).

818. H.FAHLENBRACH, "Magnetic Properties of Amorphous Metallic Materials", Elektro. & Dtsch. Elektrohandwerk, 52 (1977) 1108-10 (in German) (Mag)(see Phys. Abstr., 80(1977) 89977).

819. A.FERT, R.ASOMOZA, I.CAMPBELL and H.JOUVE, "Contribution of Magnetic Ordering to the Resistivity of Amorphous Alloys", C.R.Acad. Sci., B285 (1977) 113-6 (in French) (E)(see Phys. Abstr., 80(1977) 89746).

820. J.L.FINNEY, "Modelling the Structures of Amorphous Metals and Alloys", Nature, 266 (1977) 309-14 (SMA).

821. J.FLECHON, F.A.KUHNAST, F.MACHIZAUD, B.AUGUIN and A.DEFRESNE, "Thermal Analysis of Amorphous Ni-P and Ni-B Alloys", J. Calorimetric Anal. Therm., 8, No. 2 (1977) 61-9 (T)(in French) (see Chem. Abstr., 89(1978) 47480j).

822. H.FUJIMORI, M.KIKUCHI, Y.OBI and T.MASUMOTO, "High Permeability Properties of Amorphous Co-Fe Base Alloys", J. Japan Inst. Metals, 41 (1977) 111-6 (in Japanese)(Mag).

823. F.E.FUJITA, T.MASUMOTO, M.KITAGUCHI and M.URA, "On the Structure of Fe-P-C Amorphous Alloys", Japan. J. Appl. Phys., 16 (1977) 1731-8 (AQS).

824. K.FUKAMICHI, M.KIKUCHI, S.ARAKAWA and T.MASUMOTO, "Invar-Type New Ferromagnetic Amorphous Fe-B Alloys", Solid State Commun., 23 (1977) 955-8 (Mag).

825. N.FUNAKOSHI, T.KANAMORI and T.MANABE, "Liquid Quenching of $Pd_{82-x}Ag_xSi_{18}$ Alloys", Japan. J. Appl. Phys., 16 (1977) 1-4 (AQS).

826. N.FUNAKOSHI, T.KANAMORI and T.MANABE, "The Crystallization Kinetics of Amorphous $Pd_{78}Ag_4Si_{18}$", Japan J. Appl. Phys., 16 (1977) 515-6 (AB).

827. P.FURRER and H.WARLIMONT, "Crystalline and Amorphous Structures in Rapidly Solidified Al-Cr Alloys", Mater. Sci. & Engg., 28 (1977) 127-37 (MCS, AS).

828. T.GESZTI, "Introduction to Metallic Glasses", Fiz. Sz., 27 (1977) 121-6 (in Hungarian) (R)(see Phys. Abstr., 80(1977) 84656).

829. B.C.GIESSEN, N.M.MADHAVA, R.J.MURPHY, R.RAY and J.SURETTE, "Sheet Production of an Amorphous Zr-Cu Alloy by Plasma Spray Quenching", Met. Trans, A8 (1977) 364-6 (PS; QT).

830. J.J.GILMAN, "Metallic Glasses - A New Technology", in Crystal Growth and Materials, eds. E.Kaldis and H.J.Scheel (North-Holland Pub. Co., 1977), pp. 728 (R).

831. E.GIRT, B.LEONTIĆ and A.KURSUMOVIĆ, "Agglomeration of Vacancies in Pure Fast-Quenched Aluminium", Fizika, 9 (1977) 73-9 (see Met. Abstr., 11(1978) 13-0392)(PM).

832. A.M.GLASS, M.E.LINES, K.NASSAU and J.W.SHIEVER, "Anomalous Dielectric Behavior and Reversible Pyroelectricity in Roller-Quenched $LiNbO_3$ and $LiTaO_3$ Glass", Appl. Phys. Letters, 31 (1977) 249-51 (E).

833. U.GONSER, M.GHAFARI and H.G.WAGNER, "Metglasses and the Bernal Model", in Proc. Internat. Conf. Mössbauer Spectroscopy, eds. D.Barb and D.Tarina (Documentation Office, Cent. Inst. Phys., Bucharest, Romania, 1977), Vol. 1, pp. 237-8 (AQS)(see Chem. Abstr., 88(1978)175322a).

834. T.GOTO, H.MURAKAMI, K.TANAKA and M.NAGANO, "Fiber Textures in Copper and Iron Filaments Produced by the Glass Coated Melt Spinning", Trans. Japan Inst. Metals, 18 (1977) 557-61 (M).

835. T.GOTO and M.NAGANO, "Mechanical Properties of Silver Filament Produced by Method of Glass-Coated Melt Spinning", Trans. Japan Inst. Metals, 18 (1977) 562-6 (M).

836. T.GOTO and M.NAGANO, "The Glass Coated Melt Spinning of Gold", Sen-i Gakkaishi, 33 (1977) 302-6 (in Japanese) (AQS, M).

837. T.GOTO and M.NAGANO, "Transmission Electron Microscopic Study of Copper Filaments Produced by the Method of Glass Coated Melt Spinning", Sen-i Gakkaishi, 33 (1977) 87-9 (in Japanese) (AQS).

838. T.GOTO, M.NAGANO and K.TANAKA, "Transmission Electron Micrographic Study of Iron Filaments Produced by the Method of Glass Coated Melt Spinning", Sen-i Gakkaishi, 33 (1977) 229-31 (in Japanese)(AQS).

839. T.GOTO, M.NAGANO and K.TANAKA, "Tensile Strength of Copper and Iron Filaments Produced by the Method of Glass-Coated Melt Spinning", Trans. Japan Inst. Metals, 18 (1977) 209-13 (M).

840. T.GOTO, M.NAGANO and N.WEHARA, "Mechanical Properties of Amorphous $Fe_{80}P_{16}C_3B_1$ Filament Produced by Glass-Coated Melt Spinning", Trans. Japan Inst. Metals, 18 (1977) 759-64 (M).

841. J.E.GRAEBNER, B.GOLDING, R.J.SHUTZ, F.S.L.HSU and H.S.CHEN, "Low Temperature Properties of a Superconducting Disordered Metal", Phys. Rev. Letters, 39 (1977) 1480-3 (S).

842. W.GROSSE-NOBIS and W.WAGNER, "Fluctuation of the Driving Field in a Tape Wound Core of Metglas 2826", J. Mag. Mag. Mater., 6 (1977) 77-9 (Mag).

843. H.J.GÜNTHERODT, "Metallic Glasses", in 'Festkorperprobleme (Advances in Solid State Physics), ed. J.Treusch, Vieweg, Braunschweig, Vol. XVII (1977) pp. 25-53 (R).

844. H.-J.GÜNTHERODT, H.U.KÜNZI, M.LIARD, M.MÜLLLER and H.RUDIN, "Electronic Structure of Non-Crystalline Metals", in Non-Crystalline Solids, ed G.H.Frischat (Trans. Tech. Publications, Aedermannsdorf, Switzerland, 1977), pp. 212-7 (AQS)(see Chem. Abstr., 87 (1977) 110080e).

845. H.-J.GÜNTHERODT, H.U.KÜNZI, M.LIARD, R.MÜLLER, R.OBERLE and H.RUDIN, "Electrical Transport in Amorphous and Liquid Transition Metal Alloys", in Liquid Metals 1976, eds. R.Evans and D.A.Greenwood (Inst. Phys., London, 1977) pp. 342-51 (PP) (see Chem. Abstr., 86(1977) 164033q).

846. A.HAMZIĆ and E.BABIĆ, "The Magnetoresistivity of Dilute AlMn Alloys", Solid State Commun., 21 (1977) 607-10 (Mag).

847. R.HARRIS, M.SHALMON and M.J.ZUCKERMANN, "Electrical Resistivity of Highly Disordered Metallic Materials at High Temperatures", J. Phys. F: Metal Phys., 7 (1977) L259-61 (E).

848. R.HARRIS and D.ZOBIN, "The Random Anisotropy Model for Amorphous Metallic Alloys: Gneralized Molecular Field Calculations", J. Phys. F: Metal Phys., 7 (1977) 337-50 (SMA).

849. R.HASEGAWA and L.E.TANNER, "Metallic Glass Resistance Thermometers", J. Appl. Phys., 48 (1977) 3211-2 (A).

850. R.HASEGAWA and L.E.TANNER, "Superconducting Properties of Be-Zr Glassy Alloys Obtained by Liquid Quenching", Phys. Rev., B16 (1977) 3925-8 (S).

851. K.HASHIMOTO, M.KASAYA, K.ASAMI and T.MASUMOTO, "Electrochemical and XPS Studies on Corrosion Behaviour of Amorphous Ni-Cr-P-B Alloys", Corrosion Engg., 26 (1977) 445-52 (C) (in Japanese) (see Chem. Abstr., 89(1978) 116305r).

852. G.HAUSCH and E.TÖROK, "Young's Modulus Anomaly and Internal Friction of an Amorphous METGLAS Alloy", in Proc. Internat. Conf. Ultrasonic Attenuation in Solids, eds, R.R.Hasiguti and N.Mikoshiba (Univ. Tokyo Press, 1977) pp. 265-8 (M)(see Chem. Abstr., 90(1979) 96605h).

853. J.P.HIRTH, "The Mechanism of Shear of an Amorphous Metal", J. Mater. Sci., 12 (1977) 2540-2 (M).

854. S.HORI, H.KITAGAWA, T.MASUTANI and A.TAKEHARA, "Structure and Phase Decomposition of Rapidly Solidified Al-Zr", J. Japan Inst. Light Metals, 27 (1977) 129-37 (in Japanese)(see Met. Abstr., 11(1978) 12-0057) (AB).

855. K.HOSELITZ, "Magnetization of Fe-Si-B Metallic Glass", Phys. Stat. Sol., (a) 44 (1977) K191-3 (Mag).

856. A.HUBERT, "Amorphous Ferromagnetic Materials Quenched from the Melt", J. Mag. Mag. Mater., 6 (1977) 38-46 (in German) (Mag)(R).

857. T.IKEDA, T.YAMASHITA and M.TAKAHASHI, "Ultrasonic Velocity in Amorphous Ferromagnetic Ribbons", Japan J. Appl. Phys., 16 (1977) 1465-6 (PP).

858. Y.INOKUTI and B.CANTOR, "The Formation of Martensite in Splat-Quenched Fe-Mn and Fe-Ni-C Alloys", J. Mater. Sci., 12 (1977) 946-58 (AQS).

859. R.S.ISKHAKOV, "Spin Wave Spectrum in a Random Model of Amorphous Ferromagnets", Fiz. Tverd. Tela, 19 (1977) 3-7 (in Russian); Sov. Phys. Solid State, 19 No. 1(1977) 1-3 (Mag)(see Phys. Abstr., 80(1977) 89918).

860. T.IVEZIC, M.OČKO, E.BABIĆ and M.STUBICAR, "Correlation Between Microhardness and Magnetic Properties in Amorphous Fe-Ni Alloys", Electrotechnica, Zagreb, No. 5-6 (1977) 387-92 (in Croatian) (M, Mag)(see Phys. Abstr., 82(1977) 11688).

861. T.JAGIELINSKI, K.I.ARAI, N.TSUYA, S.OHNUMA and T.MASUMOTO, "Saturation Magnetostriction and Volume Magnetostriction of Fe-Ni-Co Amorphous Ribbons", IEEE Trans. Mag.,MAG-13 (1977) 1553-5 (Mag).

862. H.JONES, "Lattice Defects in Splat-Quenched Metals", in Vacancies '76, eds. R.E.Smallman and J.E.Harris (The Metals Society, London, 1977) pp.175-84 (R).

863. M.KACZOROWSKI, B.DABROWSKI and H.MATYJA, "Crystallization of Amorphous Tellurium Obtained by Splat Cooling ", Mater. Sci. & Engg., 29 (1977) 155-60 (AS, AB).

864. M.KACZOROWSKI, B.DABROWSKI and H.MATYJA, "Crystallization of Amorphous Te-Pb Alloys Obtained by Splat Cooling", Mater. Sci. & Engg., 29 (1977) 189-92 (AS, AB).

865. K.KAMIGAKI and S.ABE, "Internal Friction in Amorphous Magnets", in Proc. Internat. Conf. Ultrasonic Attenuation in Solids, eds. R.R.Hasiguti and N.Mikoshiba (Univ. Tokyo Press, 1977), pp. 273-6 (M)(see Chem. Abstr., 90(1979) 96607k).

866. G.KAMPF and W.BUCKEL, "Specific Heat of Amorphous and Polycrystalline Lead-Bismuth Films", Z. Physik, B27 (1977) 315-9 (T).

867. S.KOBE and A.R.FERCHMIN, "Amorphous Magnetism and Magnetic Materials: Bibliography 1950-1976", J. Mater. Sci., 12 (1977) 1713-49 (R).

868. A.M.KOROTKIKH and A.A.SHPUNT, "Criteria for Amorphous Solid Phase Formation During Cooling of the Melt", Zhur. Fiz. Khim., 51 (1977) 697-700 (in Russian)(PM)(see Chem. Abstr., 86(1977) 195341t).

869. L.KRAUS and J.SCHNEIDER, "Magnetostriction of Amorphous $(Fe_{1-x}Ni_x)_{80}P_{10}B_{10}$ Alloys", Phys. Stat. Sol., (a) 39 (1977) K161-3 (Mag).

870. H.KRONMÜLLER and H.GRIMM, "High-Field Susceptibility and the Spin-Wave Spectrum of $Fe_{40}Ni_{40}P_{14}B_6$ Alloys", J. Mag. Mag. Mater., 6 (1977) 57-60 (Mag).

871. H.KRONMÜLLER, R.SCHÄFER and G.SCHROEDER, "Investigation of the Domain Structure of Amorphous $Fe_{40}Ni_{40}P_{14}B_6$ Alloys", J. Mag. Mag. Mater., 6 (1977) 61-4 (Mag).

872. H.KRONMÜLLER and J.ULNER, "Micromagnetic Theory of Amorphous Ferromagnets", J. Mag. Mag. Mater., 6 (1977) 52-6 (Mag).

873. D.KUNSTELJ and K.KRANJC, "Precipitation Hardening in a Metastable Liquid-Quenched Al-Ni Alloy", Wissen Zeit. Pedagog. Hochschule, N.K.Krupskaja, 15(1977) 14 (AB).

874. M.LASOCKA and H.MATYJA, "Mechanism of Crystallization in Glassy Eutectic Te-Ge Alloy", in Non-Crystalline Solids, ed. G.H. Frischat (Trans Tech Publications, Aedermannsdorf, Switzerland, 1977) pp. 184-9 (AB)(see Met. Abstr., 11(1978) 12-0095).

875. B.G.LEWIS and H.A.DAVIES, "The Formation of Amorphous Metallic Phases by Continius Cooling from the Liquid State", in Liquid Metals 1976, eds. R.Evans and D.A.Greenwood (Inst. Phys., London, 1977) pp. 274-82 (AS).

876. B.G.LEWIS and H.A.DAVIES, "Influence of Composition and Cooling Rate on the Satability and Formation of Metallic Glasses", in Proc. Internat. Symp. Structure of Non-Crystalline Materials, ed. P.H.Gaskell (Taylor & Francis, London, 1977) pp. 89-93 (R, AS).

877. L.LEY and J.D.RILEY, "Formation of a Protective Silica Overlayer During the Oxidation of an Amorphous $Pd_{0.81}Si_{0.19}$ Alloy", in Proc. 7th Internat. Vac. Cong., eds. R.Dobrozemsky, F.Ruedenauer and F.P.Viehboeck (Vienna, Austria, 1977), Vol. 3, pp. 2031-3 (C)(see Chem. Abstr., 88(1978)140545y).

878. H.H.LIEBERMANN, C.D.GRAHAM, Jr. and P.J.FLANDERS, "Changes in Curie Temperature, Physical Dimensions and Magnetic Anisotropy during Annealing of Amorphous Magnetic Alloys", IEEE Trans. Mag., MAG-13 (1977) 1541-3 (Mag).

879. J.LIVAGE, D.BALLUTAND, M.C.LEROY, M.MICHAUD and C.R'KHA, "Electron Localization in Doped Amorphous V_2O_5", in Amorphous and Liquid Semi-Conductors (Proc. 7th Internat. Conf.), ed. W.E.Spear (Centre for Consultancy and Liaison, Univ. Edinburgh, 1977) pp. 654-8 (E).

880. K.LO, J.BEVK and D.TURNBULL, "Critical Currents in Liquid-Quenched Nb_3Al", J. Appl. Phys., 48 (1977) 2597-2600 (S).

881. A.E.LORD, Jr. and P.M.ANDERSON III, "Acoustic Emissions Generated during Viscous Flow of Metallic Glasses", Letters Appl. Engg. Sci., 5 (1977) 335-9 (PP)(see Phys. Abstr., 81 (1978) 39853).

882. A.E.LORD, Jr. and P.M.ANDERSON III, "Acoustic Emiss ions Generated During Viscous Flow of Metallic Glasses", Letters Appl. & Eng. Sci., 5(1977) 489-94 (PP)(see Phys. Abstr., 81 (1978) 42915).

883. F.E.LUBORSKY, "Magnetism in Amorphous Alloys", Nature, 267 (1977) 583-4 (Mag).

884. F.E.LUBORSKY, "Crystallization of Some Fe-Ni Metallic Glasses", Mater. Sci. & Engg., 28 (1977) 139-44 (AB).

885. F.E.LUBORSKY, D.CHEN and R.E.TOMPKINS, "Electronic Transformer Using Amorphous Material", IEEE Trans. Mag., MAG-13 (1977) 1224-6 (A).

886. F.E.LUBORSKY and J.L.WALTER, "Kinetics of Reorientation of the Induced Anisotropy in Amorphous $Fe_{40}Ni_{40}B_{20}$", Mater. Sci. & Engg., 28 (1977) 77-80 (Mag).

887. F.E.LUBORSKY and J.L.WALTER, "Magnetic Anneal Anisotropy in Amorphous Alloys", IEEE Trans. Mag., MAG-13 (1977) 953-6 (Mag).

888. F.E.LUBORSKY and J.L.WALTER, "Magnetically Induced Anisotropy in Amorphous Alloys of Fe-Ni-P-B", IEEE Trans. Mag., MAG-13 (1977) 1635-8 (Mag).

889. J.D.MACKENZIE, "New Applications of Glass", J. Non-Cryst. Solids, 26 (1977) 456-81 (A).

890. J.U.MADSEN, R.M.J.COTTERILL, F.KRAGH, K.BORGGREEN, I.HANSSON, J.K.KRISTENSEN and W.D.KRISTENSEN, "Morphology and State of Structures in Amorphous $Pd_{80}Si_{20}$ Fracture Surfaces", Scand. J. Metallurgy, 6 (1977) 15-7 (M).

891. M.L.MALHOTRA, "Electron Microscopy of Gold-Containing Ag_3Sn Splat-Cooled Ribbons", Metallography, 10 (1977) 385-98 (AQS).

892. M.A.MARCUS, "Grain Boundary Diffusion in Partially Devitrified Glasses: A New Percolation Problem", Scripta Met., 11 (1977) 831-4 (AB).

893. Ž.MAROHNIĆ, E.BABIĆ and D.PAVUNA, "Galvanomagnetic Effects in $Fe_{40}Ni_{40}B_{20}$", Phys. Letters, A63 (1977) 348-50 (Mag).

894. Ž.MAROHNIĆ, J.IVKOV and E.BABIĆ, "Hall Effect in Amorphous $Fe_xNi_{80-x}P_{14}B_6$ Alloys", Elektrotechnika, Zagreb, No. 5-6 (1977) 415-9 (in Croatian) (Mag)(see Phys. Abstr., 82(1979) 11102).

895. J.MARTI and D.I.PAUL, "The Anisotropy Dependence of Ferromagnetic Domain Wall Widths in Crystalline and Amorphous Metals", J. Appl. Phys., 48 (1977) 4678-86 (Mag).

896. N.I.MARZWELL, "Magnetic and Electrical Properties of Amorphous Pd-Co-P Alloys", J. Mag. Mag. Mater., 5 (1977) 67-77 (E, Mag).

897. T.MASUMOTO, "Amorphous Metals as Future Magnetic Materials", Electronic Parts and Materials, 16, No. 10 (1977) 95-7 (in Japanese) (R, Mag).

898. T.MASUMOTO and H.FUJIMORI, "Development of New Type Amorphous Soft Magnetic Alloys and Their Applications", Chemistry & Chemical Industry, 30 (1977) 723-6 (in Japanese) (R, Mag).

899. T.MASUMOTO, A.INOUE and H.M.KIMURA, "Crystallization Process of Iron-, Nickel- and Cobalt-Based Amorphous Alloys Containing Silicon and Boron", J. Japan Inst. Metals, 41 (1977) 730-7 (in Japanese) (AB)(see Met. Abstr., 11(1978) 12-0254).

900. J.R.MATEY and A.C.ANDERSON, "Phonon Transport in Glassy Metals Below 100 K", Phys. Rev., B16 (1977) 3406-10 (PP).

901. J.R.MATEY and A.C.ANDERSON, "Low Temperature Thermal Conductivity of an Amorphous Pd-Si Alloy", J. Non-Cryst. Solids, 23 (1977) 129-37 (T).

902. L.V.MEISEL and P.J.COTE, "Application of the Extended Ziman Theory to Amorphous Ni-P Alloys", Phys. Rev., B15 (1977) 2970-3 (ED; E).

903. M.MICHAUD, P.PINEAU, J.LIVAGE and R.COLLONGUES, "Electronic Mobility in Doped Amorphous V_2O_5 Semiconductors", Rev. Appl. Physique, 12 (1977) 715-9 (in French) (E).

904. O.MILAT and D.DUŽEVIĆ, "Splat Quenching Device for Rapid Quenching High Melting Temperature Metals and Metallic Alloys", Elektrotehnika (Zagreb), No. 5-6 (1977) 420-3 (in Croatian) (QT)(see Phys. Abstr., 82(1979) 11639).

905. M.MILKOVIC, F.E.LUBORSKY, D.CHEN and R.E.TOMPKINS, "Electronic Transformer Using Amorphous Materials", IEEE Trans. Mag., MAG-13 (1977) 1224-6 (A).

906. I.S.MIROSHNICHENKO and V.F.BASHEV, "Formation of Amorphous Phases in Fe-B and Fe-C Alloys During Rapid Quenching From Melt", Izv. Akad. Nauk SSSR, Met., No. 3 (1977) 206-7 (in Russian); Russian Met., No. 3 (1977) 178-9 (AS)(see Met. Abstr., 10(1977) 11-0813).

907. H.MIYAJIMA, "Physics of Amorphous Magnetism", Solid State Phys., 12(1977) 672-80 (in Japanese) (R, Mag)(see Phys. Abstr., 81 (1978) 52259).

908. R.MOHS and U.KÖNIG, "Magnetic Properties of the Amorphous Metal Alloy $Fe_{40}Ni_{40}P_{14}B_6$", J. Mag. Mag. Mater., 6 (1977) 84-6 (Mag).

909. B.MONTEIL, J.C.BERNIER and A.REVCOLEVSCHI, "Magnetic Properties of New Phases Obtained by Rapid Quenching from High Temperature", Mater. Res. Bull., 12 (1977) 235-40 (in French) (AQS, AS, Mag).

910. H.A.MOOK and C.C.TSUEI, "Magnetic Excitations in Amorphous $Fe_{75}P_{15}C_{10}$", Phys. Rev., B16 (1977) 2184-90 (Mag).

911. K.MUKASA and K.MAEDA, "Internal Friction in an Amorphous Ferromagnetic Alloy", in Proc. Internat. Conf. Ultrasonic Attenuation in Solids, eds. R.R.Hasiguti and N.Mikoshiba (Univ. Tokyo Press, 1977) pp. 269-72 (M)(see Chem. Abstr., 90 (1979) 96606j).

912. R.MÜLLER, H.J.GÜNTHERODT, H.U.KÜNZI, M.LIARD, M.MÜLLER and H.RUDIN, "Electronic Structure of Non-Crystalline Metals", in Non-Crystalline Solids, ed. G.H.Frischat (Trans Tech. Publications, Switzerland, 1977) pp. 212-7 (E).

913. A.NAGATA, H.HIRAYAMA, K.NOTO and O.IZUMI, "Superconductivity in a Splat-Quenched and Aged Copper-0.4 at.% Nb Alloy", J. Japan Inst. Metals, 41 (1977) 578-83 (in Japanese) (S)(see Met. Abstr., 11(1978) 33-0789).

914. S.R.NAGEL, "Temperature Dependence of the Resistivity in Metallic Glasses", Phys. Rev., B16 (1977) 1694-8 (E).

915. S.R.NAGEL and J.TAUC, "Correlations in Binary Liquid and Glassy Metals", Solid State Commun., 22 (1977) 129-32 (PM).

916. S.R.NAEL and J.TAUC, "Influence of Electronic Structure on Glass Forming Ability of Alloys", in Liquid Metals 1976, eds. R.Evans and D.A.Greenwood (Inst. Phys., London, 1977), pp. 283-7 (AS).

917. S.R.NAGEL, J.TAUC and B.C.GIESSEN, "Photoemission Study of Nb-Ni Glasses", Solid State Commun., 22 (1977) 471-4 (PP).

918. K.NARITA, H.FUKUNAGA and J.YAMASAKI, "Effects of Metalloid Content on Curie Temperature and Magnetic Moment of Amorphous Fe-Si-B Alloys", Japan. J. Appl. Phys., 16 (1977) 2063-4 (Mag).

919. K.NARITA, J.YAMASAKI and H.FUKUNAGA, "Dependence on Metalloid Content for Magnetic Properties of Fe-Si-B Metallic Glasses", IEEE Trans. Mag., MAG-13 (1977) 1544-6 (Mag).

920. R.C.O'HANDLEY, "Magnetostriction of Ferromagnetic Metallic Glasses", Solid State Commun., 21 (1977) 1119-22 (Mag).

921. R.C.O'HANDLEY, "Temperature Dependence of Magnetostriction in $Fe_{80}B_{20}$ Glass", Solid State Commun., 22 (1977) 485-8 (Mag).

922. R.C.O'HANDLEY, R.HASEGAWA, R.RAY and C.-P.CHOU, "Magnetic Properties of $TM_{80}P_{20}$ Glasses", J. Appl. Phys., 48 (1977) 2095 only (Mag).

923. T.OHASHI and R.ICHIKAWA, "On Rapidly Solidified and Annealed Structures of Al-0.5∿2.5% Ti Alloys", J. Japan Inst. Light Metals, 27 (1977) 105-12 (in Japanese) (AB).

924. I.OHDOMARI, "Formation of Amorphous Layer by Ion Implantation", Oyo Buturi, 46 (1977) 793-802 (in Japanese) (R)(see Phys. Abstr., 81(1978) 22746).

925. F.J.OHKAWA and K.YOSIDA, "Theory of the Electrical Resistivity of Amorphous Metals", J. Phys. Soc. Japan, 43 (1977) 1545-9 (E).

926. S.OHNUMA, K.WATANABE and T.MASUMOTO, "Low Coercivity and Zero Magnetostriction of Amorphous Fe-Co-Ni System Alloys", Phys. Stat. Sol., (a) 42 (1977) K151-4 (Mag).

927. D.G.ONN, W.D.JOHNSON, P.F.GLEESON, T.A.DONNELLY, T.EGAMI and H.H.LIEBERMANN, "Low-Temperature Heat Capacity of the Metallic Glasses $Fe_{80}P_{14}B_6$, $Fe_{40}Ni_{40}P_{14}B_6$ and $Ni_{80}P_{14}B_6$", J. Phys. C: Solid State Phys., 10 (1977) L639-42 (T).

928. D.PAVUNA, "Metallic Glasses-Definition, Techniques of Preparation and Potential Applications", Elektrotehnica (Zagreb), No. 5-6 (1977) 424-9 (in Croatian) (R)(see Phys. Abstr., 82 (1979) 10800).

929. F.PFEIFER and W.BEHNKE, "Magnetic Properties of Amorphous Alloy Toroids Based on Iron and Nickel", J. Mag. Mag. Mater., 6 (1977) 80-3 (in German) (Mag).

930. J.M.POATE, J.A.BORDERS, A.G.CULLIS and J.K.HIRVONEN, "Ion Implantation as an Ultrafast Quenching Technique for Metastable Alloy Production: The Ag-Cu System", Appl. Phys. Letters, 30 (1977) 365-8 (II; QT).

931. S.J.POON and J.DURAND, "The Ruderman-Kittel-Kasuya-Yosida Interaction in Amorphous $La_{80}Au_{20}$ Alloys with Dilute Gd Impurities", Solid State Commun., 21 (1977) 793-5 (Mag).

932. S.J.POON and J.DURAND, "Superconductivity and Spin-Glass Regime in Amorphous $La_{80}Au_{20}$ Alloys Doped with Gd", Solid State Commun., 21 (1977) 999-1002 (S).

933. S.J.POON and J.DURAND, "Remanent Magnetization of Amorphous La-Gd-Au Alloys with High Gd Content", Commun. on Physics, 2 (1977) 87-93 (Mag).

934. S.J.POON and J.DURAND, "Critical Phenomena and Magnetic Properties of an Amorphous Ferromagnet: Gd-Au", Phys. Rev., B16 (1977) 316-30 (Mag).

935. S.J.POON, J.DURAND and M.YUNG, "Magnetic Field Dependence of Resistivity Minima in Amorphous La-Gd-Au Alloys with High Gd Content", Solid State Commun., 22 (1977) 475-9 (E, Mag).

936. K.RAJ, J.DURAND, J.I.BUDNICK, C.C.TSUEI and S.SKALSKI, "^{31}P NMR in the Ferromagnetic State of Amorphous $Fe_{75}P_{15}C_{10}$", Solid State Commun., 24 (1977) 189-92 (Mag).

937. P.RAMACHADRARAO, B.CANTOR and R.W.CAHN, "Viscous Behaviour of Undercooled Metallic Melts", J. Non-Cryst. Solids, 24 (1977) 109-20 (PP).

938. P.RAMACHANDRARAO, B.CANTOR and R.W.CAHN, "Free Volume Theories of the Glass Transition and the Special Case of Metallic Glasses", J. Mater. Sci., 12 (1977) 2488-2502 (T).

939. O.RAPP, S.M.BHAGAT and Ch.JOHANNESON, "Measurements of the Electrical Resistivity of Amorphous Ferromagnets", Solid State Commun., 21 (1977) 83-7 (E).

940. R.RAY, R.HASEGAWA, C.-P.CHOU and L.A.DAVIS, "Iron-Boron Glasses: Density, Mechanical and Thermal Behaviour", Scripta Met., 11 (1977) 973-8 (PP, M, T).

941. M.D.RECHTIN, J.B.VANDERSANDE and P.M.BALDO, "Annual Meet. EMSA, 35 (1977) 42-3 (AB).

942. J.F.SADOC and J.DIXMIER, "Structural Investigation of Amorphous Binary Alloys by X-Ray and Neutron Scattering", in Proc. Internat. Symp. Structure of Non-Crystalline Materials, ed. P.H.Gaskell (Taylor & Francis, London, 1977) pp. 85-8 (AQS) (see Chem. Abstr., 89(1978) 204423n).

943. E.M.SAVITSKY and Yu.V.JEFIMOV, "The Effects of Rapid Quenching of Superconductors", Doklady Akad. Nauk SSSR, 232 (1977) 1058-61 (in Russian) (S); Sov. Phys. -Doklady, 22 (1977) 91-3 (see Chem. Abstr., 86 (1977) 144304a).

944. E.M.SAVITSKY, Yu.V.JEFIMOV and T.M.FROLOVA, "Ultrafast Quenching of the Compound V_3Sn", Doklady Akad. Nauk SSSR, 236 (1977) 582-4 (in Russian); Sov. Phys. Doklady, 22 (1977) 523-5 (S).

945. E.M.SAVITSKY, Ch.J.RAUB, Yu.V.JEFIMOV, H.R.KHAN and G.G.MUCHIN "Effect of Rapid Quenching on the Phase Transition of Alloys", Metall, 31 (1977) 1091-5 (see Met. Abstr., 11(1978) 11-0332) (AQS).

946. E.M.SAVITSKY, E.SAUR, Ch.J.RAUB and Yu.V.JEFIMOV, "Fast Quenching of A-15 Phases", Z. Metallkde., 68 (1977) 128-34 (in German) (S).

947. A.S.SCHAAFSMA, W.VENHUIZEN and F.VAN DER WOUDE, "Isothermal Crystallization of Metals Iron-Boron ($Fe_{80}B_{20}$)", in Proc. Internat. Conf. Mössbauer Spectroscopy, eds. D.Barb and D.Tarina (Documentation Office, Cent. Inst. Phys., Bucharest, Romania, 1977), Vol. 1, pp. 239-40 (AB)(see Chem. Abstr., 88 (1978) 113446k).

948. J.SCHNEIDER, K.ZAVETA, A.HANDSTEIN, R.HESSKE and W.HAUBENREISSER, "Magnetization in Amorphous Nickel-Iron-Base Alloys", Physica, 91B+C (1977) 185-9 (Mag).

949. L.J.SCHOWALTER, M.B.SALAMON, C.C.TSUEI and R.A.CRAVEN, "The Critical Specific Heat of a Glassy Ferromagnet", Solid State Commun., 24 (1977) 525-9 (T).

950. M.G.SCOTT, "Vapour Quenching of Cu-Zr Alloys", Mater. Sci. & Engg., 30 (1977) 219-22 (VD; AQS).

951. M.G.SCOTT and P.RAMACHANDRARAO, "Kinetics of Crystallization of Fe-P-C Glass", Mater. Sci. & Engg., 29 (1977) 137-44 (AB).

952. W.H.SHULL and D.G.NAUGLE, "Specific Heats of Superconducting Splat-Cooled 'Amorphous' $La_{1-x}Au_x$ Alloys", Phys. Rev. Letters, 39 (1977) 1580-3 (T).

953. J.SITEK, J.CIRAK, M.PREJSA and M.HUCL "Mössbauer Spectroscopy of the Amorphous Alloy $Fe_{75}P_{15}C_{10}$", in Proc. Internat. Conf. Mössbauer Spectroscopy, eds. D.Barb and D.Tarina (Documentation Office, Cent. Inst. Phys., Bucharest, Romania, 1977), Vol. 1, pp. 233-4 (AQS)(see Chem. Abstr., 88(1978) 112993z).

954. F.SPAEPEN, "A Microscopic Mechanism for Steady State Inhomogeneous Flow in Metallic Glasses", Acta Met., 25 (1977) 407-15 (M).

955. M.STUBIČAR, E.BABIĆ, D.SUBASIC, D.PAVUNA and Ž.MAROHNIĆ, "Phase Transformations during Isochronal Annealing of $Fe_{40}Ni_{40}$-B_{20} Glass", Phys. Stat. Sol., (a) 44 (1977) 339-44 (AB).

956. K.SUZUKI and T.FUKUNAGA, "Time of Flight Pulsed Neutron Diffraction of Co-P Amorphous Alloy made by Electrodeposition", Res.Rep. Nucl. Sci. Tohoku Univ., 10 (1977) 209-14 (ED; SMA)(see Chem. Abstr., 90(1979) 113134y).

957. T.SUZUKI, S.SAITO and W.ARAKAWA, "The Effect of High Pressure on Crystallization of Splat Cooled V_2O5", J. Non-Cryst. Solids, 24 (1977) 355-60 (AB).

958. L.TAKAĆS, "Mössbauer Investigation of the Magnetic Anisotropy and Electron Structure of a Metallic Glass", Solid State Commun., 21 (1977) 611-3 (Mag).

959. M.TAKAHASHI, "A General View on Ferromagnetic Amorphous Materials (Magnetism and Structure of Amorphous $Fe_{80}P_{13}C_7$)", Kumsok Hakhoe Chi, 15 (1977) 319-23 (R)(see Chem. Abstr., 87 (1977) 176500y).

960. M.TAKAHASHI, "Ferromagnetic Amorphous Materials", Taehan Kumsok Hakhoe Chi, 15 (1977) 319-23 (in Japanese) (R, Mag)(see Chem. Abstr., 88(1978) 57573K).

961. M.TAKAHASHI and S.ISHIO, "Uniaxial Magnetic Anisotropy of an Amorphous $Fe_{82}P_{18}$ Alloy", Japan. J. Appl. Phys., 16 (1977) 2273-4 (Mag).

962. M.TAKAHASHI and C.O.KIM, "Temperature Dependence of the Magnetic Anisotropy in Amorphous $Fe_{81}B_{19}$ Alloy", Japan. J. Appl. Phys., 16 (1977) 2061-2 (Mag).

963. M.TAKAHASHI and M.KOSHIMURA, "Temperature Dependence of the Saturation Magnetization in Amorphous Fe-B Alloys", Japan. J. Appl. Phys., 16 (1977) 1711-2 (Mag).

964. M.TAKAHASHI and M.KOSHIMURA, "Temperature Dependence of the Saturation Magnetization in Amorphous Fe-P Alloys", Japan. J. Appl. Phys., 16 (1977) 2269-70 (Mag).

965. M.TAKAHASHI, T.SUZUKI and T.MIYAZAKI, "Magnetic Domains of an Amorphous $Fe_{80}P_{13}C_7$ Alloy", Japan. J. Appl. Phys., 16 (1977) 521-2 (Mag).

966. L.E.TANNER and R.RAY, "Physical Properties of $Ti_{50}Be_{40}Zr_{10}$ Glass", Scripta Met., 11 (1977) 783-9 (PP).

967. N.TEOH, W.TEOH and S.ARAJS, "Electrical Resistivity of Amorphous Metallic Glasses $Fe_{80}B_{20}$ and $Fe_{78}Mo_2B_{20}$ between 78 and 1000 K", J. Phys. Chem. Solids, 38 (1977) 919-21 (E).

968. C.C.TSUEI, W.L.JOHNSON, R.B.LAIBOWITZ and J.M.VIGGIANO, "The Ratio of Energy Gap to Transition Temperature in Amorphous Superconductors", Solid State Commun., 24 (1977) 615-8 (Sp, VD; S).

969. N.TSUYA and K.I.ARAI, "Theory of Giant ΔE Effect in Magnetostrictive Amorphous Ribbons", IEEE Trans. Mag., MAG-13 (1977) 1547-9 (Mag).

970. N.TSUYA and K.I.ARAI, "New Magnetostrictive Materials and Devices", Oyo Buturi, 46 (1977) 813-6 (in Japanese) (R)(see Phys. Abstr., 81(1978) 23316).

971. D.TURNBULL, "On the Gram-Atomic Volumes of Metal-Metalloid Glass Forming Alloys", Scripta Met., 11 (1977) 1131-6 (PP).

972. J.VANDER BORST, F.J.A.DEN BROEDER and T.SCHEFFERS, "The Effect of Metalloid Composition on Some Magnetic and Mechanical Properties of $Co_{75}Si_{25-x}B_x$ Glasses", J. Appl. Phys., 48 (1977) 1724-6 (M, Mag).

973. A.M.VAN DIEPEN and F.J.A.DEN BROEDER, "Effect of Thermal Stress on the Easy Axis of Amorphous $Fe_{40}Ni_{40}P_{14}B_6$", J. Appl. Phys., 48 (1977) 3165-6 (Mag).

974. B.VASVARI, "Electrical Resistivity of Amorphous Transition Metals and Their Alloys", Rep. KFKI-77-81, Hungarian Acad. Sci., Budapest (October 1977), 12pp. (E)(see Phys. Abstr., 81(1978) 60103).

975. J.L.WALTER, D.G.LEGRAND and F.E.LUBORSKY, "Small-Angle X-Ray Scattering from Amorphous Alloy $Fe_{40}Ni_{40}P_{14}B_6$", Mater. Sci. & Engg., 29 (1977) 161-7 (AQS).

976. J.L.WALTER, P.RAO, E.F.KOCH and S.F.BARTRAM, "Microstructural Study of Crystallization of Amorphous Alloy $Ni_{40}Fe_{40}P_{14}B_6$", Met. Trans., A8 (1977) 1141-8 (AB).

977. R.WANG and M.D.MERZ, "Polymorphic Bonding and Thermal Stability of Elemental Non-Crystalline Solids", Phys. Stat. Sol., (a) 39 (1977) 697-703 (T).

978. Y.WASEDA, "On the Structural Model and Short-Range Order in Amorphous Metals", Solid State Phys., 12 (1977) 181-5 (in Japanese)(R)(see Phys. Abstr., 80(1977)84654).

979. Y.WASEDA, T.MASUMOTO and S.TAMAKI, "A Comparison Between the Structure of Amorphous Ni-P and Cu-Zr Alloys", in Liquid Metals 1976, eds. R.Evans and D.A.Greenwood (Inst. Phys., London) 1977) pp. 268-73 (AQS).

980. Y.WASEDA, T.MASUMOTO and S.TOMIZAWA, "Structural Study of Amorphous Cu-Zr Alloys", Canad. Met. Quart., 16 (1977) 142-6 (AQS).

981. Y.WASEDA, H.OKAZAKI and T.MASUMOTO, "Current Views on the Structure and Crystallization of Metallic Glasses", J. Mater. Sci., 12 (1977) 1927-49 (R).

982. Y.WASEDA, H.OKAZAKI and T.MASUMOTO, "A Comparison Between the Structures of Oxide Glass (SiO_2) and Metallic Glass (Fe-P)", in Proc. Internat. Symp. Structure of Non-Crystalline Materials, ed. P.H.Gaskell (Taylor & Francis, London, 1977) pp. 95-7 (AQS).

983. Y.WASEDA and J.M.TOGURI, "A Review of the Structure and Properties of Metallic Glasses", Canad. Met. Quart., 16 (1977) 133-41 (R).

984. Y.WASEDA and J.G.WRIGHT, "The Resistivity of Amorphous Transition Elements", Phys. Stat. Sol., (b) 81 (1977) K37-40 (E).

985. J.D.WEEKS, "One and Two-Component Hard Sphere Models for the Structure of Metallic Glasses. An Integral Equation Approach", Phil. Mag., 35 (1977) 1345-63 (SMA).

986. H.WIESNER and J.SCHNEIDER, "Stability and Phase Transformations in Metastable Metallic Phases", in Strikturunters. Nichkrist. Partiellkrist. Stoffen, Mater. Arbeitstag, 1977, pp. 33-8 (in German) (R)(see Chem. Abstr., 90(1979) 113048y).

987. D.B.WILLIAMS and J.W.EDINGTON, "Microstructural Characteristics of Splat-Quenched Al-Cu Alloys", J. Mater. Sci., 12 (1977) 126-30 (AQS).

988. J.G.WRIGHT, "Transition Metals in the Amorphous State", in Liquid Metals 1976, eds. R.Evans and D.A.Greenwood (Inst. Phys., London, 1977), pp. 257-61 (AS)(see Chem. Abstr., 86(1977) 177426v).

989. A.A.YAKUNIN, I.I.OSIPOV, V.I.TKACH and A.B.LYSENKO, "Influence of Quenching from the Melt and Heat Treatment on the Structure and Properties of Al-Mn Alloys", Fiz. Met. Metalloved., 43 (1977) 140-4 (in Russian); Phys. Met. Metalloved., 43, No. 1 (1977) 120-4 (AQS, AB).

990. T.YAMAGUCHI and K.NARITA, "Magnetothermal Characteristics of Amorphous Fe-Si-B Powder Produced by Spark Erosion", IEEE Trans. Mag., MAG-13 (1977) 1621-3 (Mag).

991. R.YAMAMOTO, H.MATSUOKA and M.DOYAMA, "Relaxed Bernal Structure as a Model of Amorphous Iron", J. Phys. F: Metal Phys., 7 (1977) L243-6 (SMA).

992. N.Y.YEDNERAL, V.A.LYAKISHEV and Yu.A.SKAKOV, "Investigation of the Structure and Phase Composition of the Alloy Fe-3.8%C-4.7%Si Quenched from the Liquid State", Fiz. Met. Metalloved., 43 (1977) 426-7 (in Russian); Phys. Met. Metallogr., 43, No. 2 (1977) 173-4 (AQS).

993. J.ZIRKE, A.TAUSEND, H.WINTER and D.WOBIG, "Infrared and Structural Studies of Amorphous TlSe", in Non-Crystalline Solids, ed. G.H.Frischat (Trans Tech Publications, Switzerland, 1977) pp. 178-83 (AQS).

AMORPHOUS MAGNETISM II, eds. R.A.Levy and R.Hasegawa (Plenum Press, New York, 1977)

994. A.AMAMOU, "Ferromagnetic and Antiferromagnetic Coupling in Amorphous $(Ni_{100-c}Mn_c)_{78}P_{14}B_8$", pp. 265-74 (Mag).

995. M.BAIBICH, R.W.COCHRANE, W.B.MUIR and J.O.STROM-OLSEN, "The Resistivity of Amorphous Ferromagnets", pp. 297-303 (E).

996. D.S.BOUDREAUX, "Structure Simulation of Transition Metal-Metalloid Glasses - II", pp. 463-8 (SMA).

997. G.S.CARGILL III, "Current Views on the Structure of Amorphous Metals II", pp. 459-61 (Abstract only)(SMA).

998. C.L.CHIEN and R.HASEGAWA, "Mössbauer Study of a Glassy $Fe_{80}B_{20}$ Ferromagnet", pp. 289-96 (Mag).

999. W.Y.CHING and C.C.LIN, "Structural Model for Amorphous Transition Metal Binary Alloys", pp. 469-77 (SMA).

1000. P.J.COTE, G.P.CAPSIMALIS and G.L.SALINGER, "The Magnetic and Electronic Properties of Amorphous Ni-P Alloys", pp. 499-511 (ED; Mag, E).

1001. J.DURAND, "Magnetic Regimes in Amorphous Ni-Fe-P-B Alloys", pp. 305-18 (Mag).

1002. J.DURAND and M.YUNG, "Electronic and Magnetic Properties of Amorphous Fe-P-B Alloys", pp. 275-88 (E, Mag).

1003. H.FUJIMORI, H.MORITA, Y.OBI and S.OHTA, "On the Magnetically Induced Anisotropy in Amorphous Ferromagnetic Alloys", pp. 393-402 (Mag).

1004. T.FUKUNAGA, T.ICHIKAWA and K.SUZUKI, "A Proposed Structural Model for Amorphous $Pd_{0.8}Si_{0.2}$ Alloy", pp. 521-8 (SMA).

1005. B.C.GIESSEN, A.CALKA, R.RAMAN and D.J.SELLMYER, "Magnetic Susceptibility of an Amorphous Non-Transition Metal Alloy: $Mg_{0.7}Zn_{0.3}$", pp. 197-205 (Mag).

1006. G.R.GRUZALSKI, J.W.WEYMOUTH, D.J.SELIMYER and B.C.GIESSEN, "Magnetism in Amorphous Zr-Cu(Fe), Zr-Cu(Gd) and Nb-Ni(Fe)", pp. 235-44 (Mag).

1007. H.-J.GÜNTHERODT, H.U.KÜNZI, M.LIARD, M.MÜLLER, R.MÜLLER and C.C.TSUEI, "The High Temperature Electronic and Magnetic Properties of Pd Alloys in Glassy, Crystalline and Liquid State", pp. 257-63 (E, Mag).

1008. W.A.HINES, L.T.KABACOFF, R.HASEGAWA and P.DUWEZ, "^{31}P Nuclear Magnetic Resonance Study in the Metallic Glass Systems $(Ni_y Pt_{1-y})_{75}P_{25}$ and $(Ni_{0.5}Pd_{0.5})_{100-x}P_x$", pp. 207-20 (Mag).

1009. F.E.LUBORSKY, "Perspective on Application of Amorphous Alloys in Magnetic Devices", pp. 345-68 (Mag, A).

1010. T.MASUMOTO, K.WATANABE, M.MITERA and S.OHNUMA, "High Magnetic Permeability Amorphous Alloys of the Fe-Ni-Si-B System", pp. 369-77 (Mag).

1011. T.MIZOGUCHI, S.VON MOLNAR, G.S.CARGILL III, T.KUDO, N.SHIOTANI and H.SEKIZAWA, "Structure and Physical Properties of an Amorphous $Cu_{57}Zr_{43}$ Alloy", pp. 513-20 (AQS, E, T, Mag).

1012. R.C.O'HANDLEY, "Magnetostriction of Metallic Glasses", pp. 379-92 (Mag).

1013. D.I.PAUL, "Application of Domain Wall Pinning Theory to Amorphous Ferromagnetic Materials", pp. 403-13 (Mag).

1014. S.J.POON and J.DURAND, "From Superconductivity to Ferromagnetism in Amorphous Gd-La-Au Alloys", pp. 245-56 (S, Mag).

1015. K.RAJ, A.AMAMOU, J.DURAND, J.I.BUDNICK and R.HASEGAWA, "NMR and Mössbauer Studies of the Amorphous System $Fe_{79}P_{21-x}B_x$", pp. 221-33 (Mag).

1016. J.A.RAYNE and R.A.LEVY, "Resistivity of Metglas Alloys from 1.5 to 800 K", pp. 319-26 (E).

1017. A.S.SCHAAFSMA and F.VAN DER WOUDE, "Transformation of Some Amorphous Fe-P-C Alloys during Isothermal Ageing", pp. 335-43 (Mag, AB).

1018. M.TAKAHASHI, M.KOSHIMURA, T.MIYAZAKI and T.SUZUKI, "Magnetism and Structure of Amorphous $Fe_{80}P_{13}C_7$ Alloy" pp. 485-98 (Mag).

1019. W.TEOH, N.TEOH and S.ARAJS, "Electrical Resistivity and Crystallization of Amorphous Metglas 2826 and Metglas 2826A", pp. 327-33 (E, AB).

1020. C.C.TSUEI, "Some Controversial Aspects of Electronic and Magnetic Interactions in Amorphous Metallic State", pp. 181-96 (E, Mag).

PHYSICA 86-88B+C (1977) (Proceedings of the International Conference on Magnetism (ICM 76) held at Amsterdam).

1021. A.BERRADA, M.F.LAPIERRE, B.LOEGEL, P.PANISSOD, C.ROBERT and J.BEILLE, "Magnetic Properties of Amorphous Ni-P Alloys", pp. 790-2 (ED; Mag).

1022. M.C.CADEVILLE, C.LERNER and J.M.FRIEDT, "Electronic Structure of Interstitial Carbon in Ferromagnetic Transition Metals Prepared by Splat Quenching", pp. 432-4 (Mag).

1023. J.F.COCHRAN, B.HEINRICH and R.BAARTMAN, "24 GHz Transmission Studies on the Amorphous Alloy Metglas 2826 from 26 to 270°C", pp. 807-9 (Mag).

1024. R.W.COCHRANE and J.O.STROM-OLSEN, "Resistance and Magneto-resistance of Amorphous Ferromagnets", pp. 779-80 (Mag).

1025. M.A.CONTINENTINO and N.RIVIER, "Spin Wave Spectrum of Amorphous Ferromagnets", pp. 793-5 (Mag).

1026. G.DIETZ and A.HÜNSELER, "Magnetic Domains in Amorphous Co-P Alloys", pp. 1337-8 (ED; Mag).

1027. F.FIORILLO, P.MAZZETTI, F.VINAI and G.P.SOARDO, "Barkhausen Noise in Amorphous Ferromagnetic Materials", pp. 803-4 (Mag).

1028. G.GRÜNER and E.BABIĆ, "Impurity Interactions in Kondo Systems", pp. 850-1 (Mag).

1029. A.HAMZIĆ and E.BABIĆ, "The Magnetoresistivity of AlMn Alloys", pp. 463-4 (Mag).

1030. R.C.JONES and G.J.YATES, "Excitations of Amorphous Ferro-magnets", pp. 785-6 (Mag).

1031. H.KRONMÜLLER and H.GRIMM, "The High-Field Susceptibility and the Micro-Structure of Amorphous Ni-Fe-P-B Alloys", pp. 805-6 (AQS, Mag).

1032. R.MALMHÄLL, K.V.RAO, G.BACKSTRÖM and S.M.BHAGAT, "Hall Effect in Metglas 2826A and 2826 - Dependence on Temperature and Pressure", pp. 796-8 (Mag).

1033. M.MILJAK and J.R.COOPER, "Single Impurity and Interaction Effects in the Magnetic Susceptibility of AlMn and AlCr Alloys", pp. 476-8 (Mag).

1034. M.MÜLLER, R.MÜLLER, M.LIARD, H.U.KÜNZI, H.-J.GÜNTHERODT, J.D.RILEY, L.LEY, G.GÜNTHERODT and C.C.TSUEI, "Magnetic Properties of Amorphous and Liquid Pd-Si Alloys", 799-800 (Mag).

1035. J.SCHNEIDER, A.HANDSTEIN, R.HESSKE and K.ZAVETA, "Magnetic Properties of Rapidly Quenched Fe-Ni Based Alloys", pp. 301-2 (Mag).

1036. N.TSUYA, K.I.ARAI and M.YAMADA, "Role of Magnetostriction on the High Frequency Magnetic Characteristics in Amorphous Ferromagnetic Ribbons", pp. 775-6 (Mag).

1037. H.WATANABE, T.MASUMOTO, M.KAMEDA, N.KAZAMA and H.YAMAUCHI, "Difference Between the Magnetic Properties of Amorphous and Crystalline Co-B-Si", pp. 801-2 (Mag).

1038. A.ZENTKO, L.POTOCKÝ, S.ULIČIANSKY and P.DUHAJ, "Magnetic Properties of Amorphous Gd-Pd-Si Alloys", pp. 777-8 (Mag).

1039. A.ZENTKOVÁ, P.DUHAJ, A.ZENTKO and T.TIMA, "Low-Field Magnetic Susceptibility of Amorphous $Co_xPd_{80-x}Si_{20}$ and $Fe_xPd_{80-x}Si_{20}$ Alloys in the Vicinity of the Magnetic Transition Temperature", pp. 787-9 (Mag).

AMORPHOUS MATERIALS ISSUE II (Sci. Rep. RITU A26 (1977)).

1040. K.FUKAMICHI, M.KIKUCHI and T.MASUMOTO, "Thermal Expansion Anomaly in Some Ferromagnetic Amorphous Alloys", pp. 225-31 (PP).

1041. H.KAYANO, T.MASUMOTO, S.TOMIZAWA and S.YAJIMA, "Neutron Irradiation Effects in Amorphous Pd-Si Alloy", pp. 240-5 (M).

1042. M.KIKUCHI, K.FUKAMICHI and T.MASUMOTO, "Young's Modulus and Delay Time Characteristics of Ferromagnetic Fe-Si-B Amorphous Alloys", pp. 232-9 (M, PP).

1043. H.M.KIMURA, T.MURATA and T.MASUMOTO, "Viscoelastic Behavior of Amorphous Metals", pp. 270-82 (M).

1044. T.MASUMOTO, "Mechanical Characteristics of Amorphous Metals", pp. 246-62 (M, R).

1045. M.NAKA, K.HASHIMOTO and T.MASUMOTO, "Effect of Chromium Addition on Corrosion Resistance of Amorphous Fe-B-C and Fe-B-Si Alloys", pp. 283-8 (C).

1046. Y.OBI, H.FUJIMORI and H.MORITA, "Hall Effect and Magnetoresistance in Ferromagnetic Amorphous Fe-Co and Fe-Ni Alloys", pp. 214-24 (Mag).

1047. S.TOMIZAWA and T.MASUMOTO, "Mechanical Properties of Amorphous Cu-Zr Alloys", pp. 263-9 (M).

1048. Y.WASEDA, H.OKAZAKI and T.MASUMOTO, "The Structure of Amorphous Iron-Phosphorus Alloys", pp. 202-7 (AQS).

1049. H.YAMAUCHI, M.KAMEDA, N.S.KAZAMA, H.WATANABE and T.MASUMOTO, "Magnetic Properties of Amorphous $(Fe_{1-x}Cr_x)$-P-C and $(Co_{1-x}Fe_x)$-B-Si Alloys", pp. 208-13 (Mag).

1978

1050. J.A.ABOAF, R.J.KOBLISKA and E.KLOKHOLM, "Properties of Transition Metal-Metalloid Ferromagnetic Thin Films", IEEE Trans. Mag., MAG-14 (1978) 941-3 (Sp; Mag).

1051. S.ADAM, E.BABIĆ and M.OČKO, "Investigations of Preferred Orientations in Zinc-based Alloys Obtained by means of Rotating Mill Device", Fizika, 10, Suppl. 2 (1978) 239-43 (AQS)(see Phys. Abstr., 82(1979) 57006).

1052. S.C.AGARWAL and H.HERMAN, "On the Question of Quenched-in Vacancies in Liquid-Quenched Al-base Alloys", Aluminium, 54 (1978) 257-60 (AQS).

1053. S.C.AGARWAL and H.HERMAN, "Phase Decomposition in Liquid-Quenched Eutectic Au-Ge Alloy", J. Mater. Sci., 13 (1978) 916-9 (AB).

1054. S.C.AGARWAL and H.HERMAN, "Early Stage Phase Separation in Liquid-Quenched Al-7 at.% Zn", J. Mater. Sci., 13 (1978) 1549-54 (AB).

1055. R.A.AKOPYAN, A.V.YEVDOKIMOV and Yu.V.DVORKIN, "Decomposition Kinetics of the Metastable Solid Solution $Ge_{0.8}(Ga\ As)_{0.2}$ Obtained by Diffusionless Crystallization", Izv. Akad. Nauk SSSR, Neorg. Mater., 14 (1978) 201-3 (in Russian); Inorganic Mater., 14 (1978) 152-4 (AB)(see Met Abstr., 11(1978) 11-0489).

1056. R.A.AKOPYAN, A.V.YEVDOKIMOV and V.I.KRASINA, "Fine Structure of Metastable Solid Solutions Prepared by Super-rapid Crystallization", Fiz. Met. Metalloved., 46 (1978) 1117-20 (in Russian); Phys. Met. Metallogr., 46, No. 5 (1978) 203-5 (AQS).

1057. R.ALBEN, J.J.BECKER and M.C.CHI, "Random Anisotropy in Amorphous Ferromagnets", J. Appl. Phys., 49 (1978) 1653-8 (Mag).

1058. A.ALI, W.A.GRANT and P.J.GRUNDY, "A Microstructural Study of Some Amorphous Transition Metal-Metalloid Surface Alloys Formed by Ion Implantation", Phil. Mag., B37 (1978) 353-76 (II; AQS).

1059. D.ALIAGA GUERRA, P.PANISSOD and J.DURAND, "^{31}P and ^{11}B NMR in Amorphous NiPB Alloys near the Paramagnetic-Ferromagnetic Transition", Solid State Commun., 28 (1978) 745-9 (Mag).

1060. P.ALLIA and F.VINAI, "Theory of Directional order and Induced Anisotropy Energy in Ferromagnetic Amorphous Systems", IEEE Trans. Mag., MAG-14 (1978) 1050-3 (Mag).

1061. A.AMAMOU and G.KRILL, "Photoemission on a Cu-Zr Glassy Metal", Solid State Commun., 28 (1978) 957-9 (PP).

1062. R.St.AMAND and B.C.GIESSEN, "On the Metastable System Mercury-Thallium", J.Less-Common Metals, 58 (1978) 161-72 (MCS, TSSE).

1063. R.St.AMAND and B.C.GIESSEN, "Easy Glass Formation in Simple Metal Alloys: Amorphous Metals Containing Ca and Sr", Scripta Met., 12 (1978) 1021-6 (AS).

1064. P.M.ANDERSON III and A.E.LORD, Jr., "Demonstration of Phase Transition and Properties of Metallic Glasses for Undergraduates", Amer. J. Phys. 46 (1978) 80-2 (see Phys. Abstr. 81(1978) 36765).

1065. T.R.ANTHONY and H.E.CLINE, "On the Uniformity of Amorphous Ribbon Formed by a Cylindrical Jet Impinging on a Flat Moving Substrate", J. Appl. Phys., 49 (1978) 829-37 (PM).

1066. C.ANTONIONE, L.BATTEZZATI, A.LUCCI, G.RIONTINO and G.VENTURELLO, "A Study of Transformation from Amorphous to Crystalline State in Fe-Ni Base Metallic Glasses", 18th National Convention AIM(Proc. Conf.) Verona, Italy, Oct. 1978, Assoc. Italianadi Metallurgia, pp. 10 (in Italian)(see Metals Abstr., 12(1978) 12-0688).

1067. C.ANTONIONE, L.BATTEZZATI, A.LUCCI, G.RIONTINO and G.VENTURELLO, "Crystallization Kinetics of Iron-Nickel Based Metallic Alloys in Non-Isothermal Conditions", J. Calorim. Therm., B18 (1978) 141-6 (in French)(AB)(see Chem. Abstr., 90(1979) 208329e).

1068. C.ANTONIONE, L.BATTEZZATI, A.LUCCI, G.RIONTINO and G.VENTURELLO, "Non-Isothermal Kinetics of Crystallization and DSC Analysis of Fe-Ni Based Metallic Glasses", Scripta Met., 12 (1978) 1011-4 (T, AB).

1069. K.I.ARAI and N.TSUYA, "Magnetomechanical Coupling and Variable Delay Characteristics by means of Giant ΔE Effect in Iron-Rich Amorphous Ribbon", J. Appl. Phys., 49 (1978) 1718-20 (Mag, M).

1070. S.ARAJS and K.W.BROWN, "Some Magnetic Properties of Glassy $Fe_{78}Mo_2B_{20}$ Between 300 and 900 K", Mater. Sci. & Engg., 34 (1978) 271-3 (Mag).

1071. S.ARAJS, C.A.MOYER and K.W.BROWN, "Determination of the Ferro-Magnetic Curie Temperature for $Fe_{40}Ni_{40}P_{14}B_6$ (Metglas 2826)", Phys. Scripta, 17 (1978) 543-5 (Mag).

1072. F.ASSMUS, "Soft Magnetic Metallic Glasses, An Introduction and Status Report", Siemens R & D Reports, 7 (1978) 118-23 (R, Mag)(see Met. Abstr., 11(1978) 33-2293).

1073. F.ASSMUS, "Recent Developments in High Permeability Alloys", Elektrotech Cas., 29 (1978) 483-91 (in Slovak) (Mag, R)(see Phys. Abstr., 82(1979) 4694).

1074. T.AYANO, H.ONO, K.KAWAMURA and K.FURUKAWA, "Structure Analysis of an Amorphous $Pd_{0.8}Si_{0.2}$ Alloy in a Dense Random Packing Model", Bussei Kenkyu, 30 (1978) 216-25 (in Japanese) (AQS) (see Chem. Abstr., 91 (1979) 100185m).

1075. E.BABIĆ, "Transport Properties of Metallic Glasses", Fizika, 10, Suppl. 2 (1978) 200-8 (see Phys. Abstr., 82(1979) 56179).

1076. E.BABIĆ, R.KRSNIK and A.HAMZIĆ, "Independence of Phonon and Spin Resistivity in Some Kondo Alloys", Solid State Commun., 25 (1978) 793-5 (E).

1077. E.BABIĆ and Ž.MAROHNIĆ, "The Disappearance of the Resistivity Minimum in Annealing Amorphous $Fe_{40}Ni_{40}B_{20}$ Alloy", J. Physique, 39 (1978) C6-948-50 (E).

1078. E.BABIĆ, Ž.MAROHNIĆ, J.R.COOPER, A.HAMZIĆ and C.RIZZUTO, "T^2 Saturation in Resistivity of Amorphous Ferromagnets Below 0.3K", J. Physique, 39 (1978) C6-946-7 (E).

1079. E.BABIĆ, Ž.MAROHNIĆ, T.IVEZIĆ and J.IVKOV, "Low Temperature Resistivities of Amorphous $Fe_xNi_{80-x}B_{20}$ Alloys", Fizika, 10, Suppl. 2 (1978) 235-8 (see Phys. Abstr., 82(1979) 56181).

1080. E.BABIĆ, Ž.MAROHNIĆ and J.IVKOV, "Low Temperature Resistivities of $Fe_xNi_{80-x}P_{14}B_6$ Metallic Glasses", Solid State Commun., 27 (1978) 441-4 (E).

1081. J.BALOGH and I.VINCZE, "Temperature Dependence of the Hyperfine Field Distribution in an Amorphous Ferromagnet", Solid State Commun., 25 (1978) 695-8 (Mag) (Report KFKI-77-80, Hungarian Acad. Sci, Budapest, October 1977).

1082. S.BARTA and R.DURNY, "Generalized Formula for Paramagnetic Susceptibility of Metallic Glasses", Czech. J. Phys., B28 (1978) 1401-4 (Mag).

1083. R.BASU, "A Computer Study of the Miscibility Limits of Pd-Cu-Si", CALPHAD (Computer Coupling Phase Diagrams & Thermochem.) 2 (1978) 785-9 (see Met. Abstr., 12(1979) 11-0457).

1084. J.J.BECKER, "A New Mechamism for Magnetic Annealing in Amorphous Metals", IEEE Trans. Mag., MAG-14 (1978) 938-40 (E, Mag).

1085. G.BELLESSA, "Frequency and Temperature Dependence of the Sound Velocity in Amorphous Materials at Low Temperatures", Phys. Rev. Letters, 40 (1978) 1456-9 (PP).

1086. L.H.BENNETT, H.E.SCHONE and P.GUSTAFSON, "Nuclear Magnetic Resonance Studies of Amorphous Ni-P Alloys", Phys. Rev., B18 (1978) 2027-30 (AQS).

1087. G.BERGMANN and P.MARQUARDT, "Resistivity of Amorphous Ferromagnetic Fe_cAu_{1-c} Alloys : Anisotropy and Field Dependence", Phys. Rev., B18 (1978) 326-37 (E).

1088. G.BERGMANN and P.MARQUARDT, "Temperature Dependence of the Resistivity in Amorphous Ferromagnetic Alloys : Spin Wave Contribution and Resistivity Minimum", Phys. Rev., B17 (1978) 1355-60 (VD; E).

1089. A.BERRADA, M.F.LAPIERRE, B.LOEGEL, P.PANISSOD and C.ROBERT, "A Study of the Magnetic and Transport Properties of Amorphous Electrodeposited Ni-P Alloys", J. Phys. F : Metal Phys., 8 (1978) 845-57 (ED; Mag, E).

1090. B.S.BERRY and W.C.PRITCHET, "Temperature Dependence of the ΔE Effect in Amorphous $Fe_{75}P_{15}C_{10}$", Solid State Comuun., 26 (1978) 827-9 (M).

1091. B.S.BERRY, W.C.PRITCHET and C.C.TSUEI, "Discovery of an Internal Friction Peak in the Metallic Glass Nb_3Ge", Phys. Rev. Letters, 41 (1978) 410-3 (M).

1092. V.N.BERZHANSKII, L.A.CHEKANOVA and N.S.CHISTYAKOV, "Ferromagnetic Resonance in Crystalline and Amorphous Magnetic Films", Fiz. Met. Metalloved., 46 (1978) 63-6 (in Russian); Phys. Met. Metallogr., 46, No. 1 (1978) 47-50 (Mag)(see Met. Abstr., 12(1979) 33-1268).

1093. S.M.BHAGAT, J.N.LLOYD and N.HEIMAN, "Spin Resonances in Amorphous Alloy Films", J. Appl. Phys., 49 (1978) 1683-5 (Mag).

1094. D.BHATTACHARYA, A.A.JOHNSON, R.K.SOREM and G.A.ANDERSEN, "The Room Temperature Dissociation of the Compound Au_3Si", Mater. Sci. & Engg., 32 (1978) 181-4 (AB).

1095. A.I.BILYAK, A.S.KOROL, A.F.NIKOLENKO, A.S.POLUKHIN and N.T.PASTUSHENKO, "A Study of the Structure of Ge-Ag Alloys", Akad. Nauk Ukr. SSR, Metallofiz., No. 71 (1978) 70-2 (AS, AB) (see Met. Abstr., 11(1978) 16-0699).

1096. R.J.BIRGENEAU, J.A.TARVIN, G.SHIRANE, E.M.GYORGY, R.C.SHERWOOD, H.S.CHEN and C.L.CHIEN, "Spin Wave Excitations and Low Temperature Magnetization in the Amorphous Metallic Ferromagnetic $(Fe_xNi_{1-x})_{75}P_{16}B_6Al_3$", Phys. Rev., B18 (1978) 2192-5 (Mag).

1097. J.L.BLACK and B.L.GYORFFY, "Interaction of the Conduction Electrons with Tunnelling States in Metallic Glasses", Phys. Rev. Letters, 41 (1978) 1595-8 (AQS).

1098. J.L.BLACK and B.L.GYORFFY, "On the Interaction Between Electrons and Tunnelling Levels in Metallic Glasses", J. Physique, 39 (1978) C6-941-2 (E).

1099. D.BLOCK, Ph.MANGIN, G.MARCHAL and C.JANOT, "High Field Magnetization Measurements in Fe_xSi_{1-x} Amorphous Alloys", Solid State Commun., 25 (1978) 555-9 (VD; Mag).

1100. G.BÖHNKE, N.CROITORIU, M.ROSENBERG and M.SOSTARICH, "Transport Properties of Ferromagnetic Metallic Glasses : Resistivity and Magnetoresistance Effect", IEEE Trans. Mag., MAG-14 (1978) 955-7 (E, Mag).

1101. V.T.BORISOV, A.I.DUKHIN and Yu.E.MATVEEV, "Principles of the Formation of Supersaturated Solid Solutions and Amorphous Alloys in Quenching from the Liquid State", Probl. Metalloved i Fiz. Met. No. 5 (1978) 4-16 (in Russian) (TSSE, AS)(see Chem. Abstr., 90(1979) 208320v).

1102. P.G.BOSWELL, "Crystallization of an $Mg_{74}Zn_{26}$ Glass", Mater. Sci. & Engg., 34 (1978) 1-5 (AB).

1103. P.G.BOSWELL, "The Spark Machining of Metallic Glasses", J. Mater. Sci., 13 (1978) 675-7 (F).

1104. P.G.BOSWELL, "Crystallization in a Phase-Separated$(Ni_5Pd_5)_{82}$-P_{18} Amorphous Alloy", Scripta Met., 12 (1978) 673-4 (AB).

1105. D.S.BOUDREAUX, "Theoretical Studies on Structural Model of Metallic Glassy Alloys", Phys. Rev., B18 (1978) 4039-47 (SMA).

1106. C.L.BRIANT, "Icosahedral Microclusters. A Possible Structural Unit in Amorphous Metals", Phys. Stat. Sol., (b) 85 (1978) 393-402 (SMA).

1107. C.L.BRIANT and J.J.BURTON, "Icosahedral Microclusters. A Possble Structural Unit in Amorphous Metals", Phys. Stat. Sol., (b) 85 (1978) 393-402 (SMA).

1108. H.A.BROOKS, "Thermal Expansion of Amorphous Metal Alloys", J. Appl. Phys., 49 (1978) 213-4 (PP).

1109. F.BROUERS, "Temperature Dependence of Electrical Resistivity in Highly Resistive Alloys", J. Physique Letters, 39 (1978) L323-6 (E).

1110. D.K.BROWN, I.NOWIK and D.I.PAUL, "A Mössbauer Study of Ferromagnetic Domain Orientation in Amorphous $Fe_{75}P_{15}C_{10}$ Foils", J. Mag. Mag. Mater., 7 (1978) 182-4 (Mag).

1111. K.H.J.BUSCHOW,"Magnetic Coupling of Rare Earth Moments in Amorphous Alloys", Solid State Commun, 27 (1978) 275-8 (Mag).

1112. K.H.J.BUSCHOW, A.M.VAN DIEPEN, N.M.BEEKMANS and J.W.M.BIESTERBOS, "Charge Transfer and Magnetic Properties in Amorphous Alloys of Fe with Y, Th or Zr", Solid State Commun., 28 (1978) 181-5 (Mag).

1113. R.W.CAHN, "Metallic Glasses", The Metallurgist & Materials Technologist, 10 (1978) 309-13 (R).

1114. R.W.CAHN, "Metallic Glasses : The Reason Why", Nature, 274 (1978) 848 (AS).

1115. R.W.CAHN, "Crystal Defects and Melting", Nature, 273 (1978) 491-2 (PM).

1116. G.S.CARGILL III, "Magnetic and Structural Aspects of Some Amorphous Systems", in Thin Film Phenomena - Interfaces and Interactions, eds. J.E.Baglin and J.M.Poate (The Electrochemical Society, Princeton, 1978) p. 221 (AQS, Mag).

1117. G.S.CARGILL III and T.MIZOGUCHI, "Dipolar Mechanisms for Magnetic Anisotropy in Amorphous Ferrimagnetic Alloys", J. Appl. Phys., 49 (1978) 1753-5 (Mag).

1118. R.CATON and A.R.SWEEDLER, "The Dependence of the Superconductivity Trnasition Temperature on Silicon Concentration in the NbAlSi Ternary System", J. Less-Common Metals, 60 (1978) 91-100 (S).

1119. M.CEDERGREN and G.BÄCKSTRÖM, "Crystallization Temperature of Amorphous $Fe_{80}B_{20}$ Under Pressure", J. Non-Cryst. Solids, 30 (1978) 69-76 (AB).

1120. W.A.CENSE and W.A.ALBERS, "On the In-Situ Growth of Aligned Composites By means of Disproportionation of Amorphous Metal Systems", J. Non-Cryst. Solids, 28 (1978) 391-401 (AB).

1121. P.H.CHANG, A.P.MALOZEMOFF, M.GRIMSDITCH, W.SENN and G.WINTERLING, "Brillouin Scattering from Ferromagnetic Metallic Glasses", Solid State Commun., 27 (1978) 617-21 (AQS).

1122. K.CHATTOPADHYAY, S.LELE and P.RAMACHANDRARAO, "On Al_2M-Type Phases in Splat-Cooled Aluminium Alloys", J. Mater. Sci., 13 (1978) 2730-1 (AQS).

1123. K.CHATTOPADHYAY and P.RAMACHANDRARO, "Phase Transitions in a Rapidly Solidified Al-2Pt Alloy", Bulletin EMSI, 2 (1978) 37-8 (AB)(see Met. Abstr., 12(1979) 11-0687).

1124. P.CHAUDHARI and D.TURNBULL, "Structure and Properties of Metallic Glasses", Science, 199 (1978) 11-21 (R).

1125. C.CHAUMONT, J.BOISSIER and J.-C.BERNIER, "Preparation by Rapid Quenching Techniques of Glasses with Lithium Ferrite", Rev. Int. Hautes Temp. & Refract., 15, No. 1 (1978) 23-32 (in French) (AS, AB)(see Phys. Abstr., 82(1979) 20019).

1126. L.A.CHEKANOVA, R.S.ISKHAKOV, R.G.KHLEBOPROS, N.S.CHISTYAKOV and G.I.FISH, "Characteristics of Crystallization of Metallic $Co_{1-x}P_x$ Glass", Fiz.Tverdogo Tela, 20 (1978) 3501-3 (in Russian); Sov. Phys. Sol. State, 20 (1978) 2026-7 (see Phys. Abstr., 82(1979) 73104).

1127. H.S.CHEN, "Method for Evaluating Viscosities of Metallic Glasses from the Rates of Thermal Transformations", J. Non-Cryst. Solids, 27 (1978) 257-63 (PP).

1128. H.S.CHEN, "Correlation between Elastic Constants and Flow Behaviour in Metallic Glasses", J. Appl. Phys., 49 (1978) 462-3 (M).

1129. H.S.CHEN, "The Influence of Structural Relaxation on the Density and Young's Modulus of Metallic Glasses", J. Appl. Phys., 49 (1978) 3289-91 (PP, M).

1130. H.S.CHEN, "Alloying Effect on the Viscous Flow in Metallic Glasses", J. Non-Cryst. Solids., 29 (1978) 223-9 (M).

1131. H.S.CHEN, "Validity of the Energy and Entropy Interfacial Tension Models Examined in view of the Stability of Metallic Glasses", Mater. Res. Bull., 13 (1978) 737-41 (T).

1132. H.S.CHEN, "The Evidence of Structural Relaxation as the Origin of Curie Temperature Aging in Metallic Glass", J. Appl. Phys., 49 (1978) 4595-7 (Mag).

1133. H.S.CHEN, L.C.KIMERLING, J.M.POATE and W.L.BROWN, "Diffusion in a Pd-Cu-Si Metallic Glass", Appl. Phys. Letters, 32 (1978) 461-3 (T).

1134. G.C.CHI, H.S.CHEN and C.E.MILLER, "The Influence of Quenching Procedures on the Kinetics of Embrittlement in a $Fe_{40}Ni_{40}B_{20}$ Metallic Glass", J. Appl. Phys., 49 (1978) 1715-7 (M).

1135. G.C.CHI and R.J.SCHUTZ, "Microstructures in Amorphous Nb_3Ge Films", Mater. Sci. & Engg., 34 (1978) 161-3 (Sp; AQS).

1136. C.L.CHIEN, "Mössbauer Study of a Binary Amorphous Ferromagnet: $Fe_{80}B_{20}$", Phys. Rev., B18 (1978) 1003-15 (AQS).

1137. C.L.CHIEN, "Hyperfine Interactions and Magnetic Ordering Temperatures of Amorphous $Fe_xPd_{80-x}Si_{20}$ with Low Fe Concentrations", Phys. Letters, A68 (1978) 394-6 (Mag).

1138. C.L.CHIEN, "Electric Field Gradient in Amorphous Magnetic Solids", Hyperfine Interactions, 4 (1978) 869-72 (Mag)(see Phys. Abstr., 81(1978) 39475).

1139. C.L.CHIEN, "$T^{3/2}$ Dependence of Hiperfine Field in Amorphous Ferromagnets", Hyperfine Interactions, 4 (1978) 866-8 (Mag) (see Phys. Abstr., 81(1978) 39474).

1140. C.L.CHIEN and R.HASEGAWA, "Mössbauer Study of Glassy Alloys $(Fe-Mo)_{80}B_{20}$", J. Appl. Phys., 49 (1978) 1721-3 (AQS, Mag).

1141. C.L.CHIEN, D.MUSSER, F.E.LUBORSKY and J.L.WALTER, "The Influence of the Metalloid Elements on the Magnetic Properties of Amorphous $Fe_{40}Ni_{40}P_yB_{20-y}$", Solid State Commun., 28 (1978) 645-9 (Mag).

1142. C.L.CHIEN, D.MUSSER, F.E.LUBORSKY and J.L.WALTER, "Hyperfine Field Distributions in Amorphous $(Fe_xNi_{1-x})_{80}P_{14}B_6$", J. Phys. F: Metal Phys., 8 (1978) 2407-15 (Mag).

1143. C.L.CHIEN, D.MUSSER, F.E.LUBORSKY and J.L.WALTER, "Mössbauer Studies of Amorphous $(Fe_xNi_{1-x})_{80}B_{20}$", J. Physique, 39 (1978) C2-62-129 (Mag).

1144. G.Y.CHIN, "New Developments in Technologically Important Magnetic Materials", J. Mag. Mag. Mater., 9 (1978) 283-98 (Mag).

1145. S.Y.CHUANG, P.K.TSENG, G.J.JAN and H.S.CHEN, "An Experimental Study of the Angular Correlation of Positron Annihilation Radiations in Electron-Irradiated Pd-Ni-Si Metallic Glass", Phys. Stat. Sol., (a) 48 (1978) K181-3 (AQS).

1146. R.CLARKE and S.R.NAGEL, "Observation of the Temperature Dependence of the Structure Factor in a Metallic Glass", Solid State Commun., 27 (1978) 215-7 (AQS).

1147. R.W.COCHRANE, "Resistivity of Amorphous Metallic Glasses", J. Physique, 39 (1978) C6-1540-4 (E, R).

1148. R.W.COCHRANE, R.HARRIS and M.J.ZUCKERMANN, "The Role of Structure in the Magnetic Properties of Amorphous Alloys", Phys. Rep., 48 (1978) 1-63 (R).

1149. R.W.COCHRANE, F.T.HEDGCOCK, B.J.KÄSTNER and W.B.MUIR, "The Resistivity of Amorphous Metallic Alloys Down to 60 mK", J. Physique, 39 (1978) C6-939-40 (E).

1150. R.W.COCHRANE, J.STROM-OLSEN, G.WILLIAMS, A.LIÉNARD and J.P.REBOUILLAT, "Magnetic and Transport Properties of Amorphous NiY and FeY", J. Appl. Phys., 49 (1978) 1677-9 (Sp; Mag).

1151. E.W.COLLINGS, C.E.MOBLEY and R.E.MARINGER, "Melt Extraction of Metallic Glass Candidate Materials", AIChE Symp. Series No. 180 - Spinning Wire from Molten Metal, 74 (1978) 102-10 (QT).

1152. P.C.COLTER, T.W.ADAIR III, D.G.NAUGLE and W.L.JOHNSON, "Hall Effect in Amorphous $La_{0.8}Ga_{0.2}$", J. Physique, 39 (1978) C6-955-6 (Mag).

1153. M.A.CONTINENTINO and N.RIVIER, "On the Resistivity of Amorphous Ferromagnets", J. Phys. F: Metal Phys., 8 (1978) 1187-1200 (E).

1154. S.M.COPLEY, D.BECK and O.ESQUIVEL, "Laser Melt Quenching and Alloying", AIP Conf. Proc., 50 (1978) 161-72 (QT)(see Phys. Abstr., 82(1979)69919).

1155. P.J.COTE and L.V.MEISEL, "Origin of Saturation Effects in Electron Transport", Phys. Rev. Letters, 40 (1978) 1586-9 (E).

1156. P.J.COTE and L.V.MEISEL, "An Alternative Mechanism to Spin Scattering for the Resistivity of Transition Metal Alloys", in Transition Metals 1977, eds. M.J.G.Lee, J.M.Perz and E.Fawcett (Inst. Phys., London, 1978) pp. 210-3 (E).

1157. R.A.CRAVEN, C.C.TSUEI and R.STEPHENS, "Specific Heat and Resistivity of an Amorphous Alloy $Fe_{34}Pd_{46}P_{20}$ Near Its Ferromagnetic Phase Transition", Phys. Rev., B17 (1978) 2206-11 (T, E).

1158. A.G.CULLIS, J.A.BORDERS, J.K.HIRVONEN and J.M.POATE, "Metastable Alloy Layers Produced by Implantation of Ag^+ and Ta^+ Ions into Cu Crystals", Phil. Mag., B37 (1978) 615-30 (II; TSSE, AS).

1159. J.A.DANTZIG and S.H.DAVIS, "Analysis for Prediction of Non-Equilibrium Phase Formation in Rapid Conduction Cooling", Mater. Sci. & Engg., 23 (1978) 199-209 (PM).

1160. N.DE CRISTOFARO and G.HENSCHEL, "Metals Brazing Foils", Weld. J., 57, No. 7 (1978) 33-8 (A)(see Chem. Abstr., 89(1978) 219464a).

1161. G.DIETZ, H.BESTGEN and J.HUNGENBERG, "About the Origin of Magnetic Anisotropy in Electrodeposited Amorphous Co-P Alloys", J. Mag. Mag. Mater., 9 (1978) 208-10 (ED; Mag).

1162. G.DIETZ and A.HUNSELER, "The Effect of Annealing on Domain Structures in Amorphous Co-P Alloys", Physica, 94B+C (1978) 223-4 (ED; Mag).

1163. G.DIETZ and K.KLETT, "Electrodeposited Amorphous Co-P Alloys with High Permeability", J. Mag. Mag. Mater., 8 (1978) 57-60 (ED; Mag).

1164. K.DOI, H.KAYANO and T.MASUMOTO, "Small-Angle Scattering from Neutron-Irradiated Amorphous $Pd_{80}Si_{20}$", J. Appl. Cryst., 11 (1978) 605 only (AQS).

1165. D.H.DOLLARD, R.H.DEE, P.W.MARTIN, B.G.TURRELL and M.J.ZUCKERMANN, "Sign and Magnitude of the Cobalt Hyperfine Field in Amorphous $Co_{80}P_{20}$ Determined by Nuclear Orientation", Solid State Commun., 25 (1978) 751-3 (Mag).

1166. E.R.DOMB and W.L.JOHNSON, "The Upper and Lower Critical Fields of High-k Amorphous Superconductors", J. Low-Temp. Phys., 33 (1978) 29-41 (S).

1167. I.W.DONALD and H.A.DAVIS, "Prediction of Glass-Forming Ability for Metallic Systems", J. Non-Cryst. Solids, 30 (1978) 77-85 (PM).

1168. P.DOUSSINEAU, P.LEGROS, A.LEVELUT and A.ROBIN, "Intensity Dependent Ultrasonic Attenuation in a Metallic Glass at Low Temperatures", J. Physique Letters, 39 (1978) L265-9 (PP).

1169. P.DOUSSINEAU, A.LEVELUT and G.BELLESSA, "Ultrasonic Attenuation in Amorphous Metallic NiP", in Proc. Internat. Conf. on Lattice Dynamics, Paris 1977 (Flammarion Sciences, Paris, 1978) pp. 462-3 (PP)(see Phys. Abstr., 81(1978) 75789).

1170. J.W.DRIJVER and S.RADELAAR, "Amorphous Metals for Flywheels", Ned. Tijdschr. Naturkund., A44 (1978) 10-14 (in Dutch) (A) (see Phys. Abstr., 81(1978)52878).

1171. J.W.DRIJVER and S.RADELAAR, "Metallic Glasses, Properties and Applications", Jaarb. Sticht. Fund. Onderz Mater Sticht. Inst. Kernphys. Onderz, 1978, pp. 167-78 (in Dutch) (R)(see Chem. Abstr., 92(1980) 9773m).

1172. J.M.DUBOIS, "An Electronic Device for Piston and Anvil Liquisol Quenching Apparatus", J. Phys. E: Sci. Instrum., 11 (1978) 76-8 (QT).

1173. J.DURAND and S.J.POON, "Crystal-Field Effects and Magnetic Interactions in Amorphous $Ce_{80}Au_{20}$", J. Appl. Phys., 49 (1978) 1702 only (Mag).

1174. J.DURAND and S.J.POON, "Spin-Glasses in Amorphous $(Ni_{100-x}Fe_x)_{79}P_{13}B_8$ Alloys", J. Physique, 39 (1978) C6-953-4 (Mag).

1175. J.DURAND, K.RAJ, S.J.POON and J.I.BUDNICK, "Local and Bulk Magnetic Properties of Amorphous Ferromagnetic Gd_2Co Obtained by Splat Cooling", IEEE Trans. Mag., MAG-14 (1978) 722-4 (Mag).

1176. P.DUWEZ, "Metallic Glasses", La Recherche, 9 (1978) 340-8 (in French) (R)(see Phys. Abstr., 81(1978) 83774).

1177. P.DUWEZ and W.L.JOHNSON, "Liquid-Quenched Amorphous Superconductors", J. Less-Common Metals, 62 (1978) 215-20 (R, S).

1178. T.EGAMI, "Structural Relaxation in Amorphous $Fe_{40}Ni_{40}P_{14}B_6$ Studied by Energy Dispersive X-Ray Diffraction", J. Mater. Sci., 13 (1978) 2587-99 (AB).

1179. T.EGAMI, "Structural Relaxation in Amorphous Alloys - Compositional Short Range Ordering", Mater. Res. Bull., 13 (1978) 557-62 (AB).

1180. T.EGAMI and S.D.DAHLGREN, "Low-Field Magnetic Properties of Sputter-Deposited Amorphous $Fe_{80}B_{20}$", J. Appl. Phys., 49 (1978) 1703-5 (Sp; Mag).

1181. T.EGAMI and T.ICHIKAWA, "Kinetics of Structural Relaxation in Amorphous Alloy Observed by X-Ray Diffraction", Mater. Sci. & Engg., 32 (1978) 293-5 (AB).

1182. E.ESPOSITO, H.EHRENREICH and C.D.GELLATT, Jr., "Electrical Transport in Transition Metal Liquids and Metallic Glasses", Phys. Rev., B18 (1978) 3913-20 (T).

1183. H.FAHLENBRACH, "Amorphous Metallic Materials and Their Magnetic Properties", Feinwerktech. Messtech., 86 (1978) 236-7 (in German) (R)(see Chem. Abstr., 89(1978) 189926h).

1184. M.FÄHNLE, "The Influence of Precipitates on the Law of Approach to Magnetic Saturation in Amorphous Ferromagnets", J. Mag. Mag. Mater., 8 (1978) 257-8 (Mag).

1185. G.FASOL, J.S.SCHILLING, C.C.TSUEI and F.ZIMMER, "Superconductivity in Amorphous $La_{78}Au_{22}$ Under High Pressure", J. Phys. F: Metal Phys., 8 (1978) L257-62 (S).

1186. R.FERRER and M.J.ZUCKERMANN, "Antiferromagnetism in Amorphous Alloys Containing Rare-Earth Atom. I. The Molecular Field Approximation", Canad. J. Phys., 56 (1978) 1098-1101 (Mag).

1187. A.FERT and I.A.CAMPBELL, "Non-axial Electric Field Gradients in Amorphous Rare-Earth Alloys", J. Phys. F: Metal Phys., 8 (1978) L57-61 (Mag).

1188. H.FISCHER, U.GONSER, R.PRESTON and H.-G.WAGNER, "Stress-Induced Magnetic Easy Axis in Amorphous $Fe_{40}Ni_{40}P_{14}B_6$", J. Mag. Mag. Mater., 9 (1978) 336-8 (Mag).

1189. J.FLECHON, F.A.KUHNAST and F.MACHIZAUD, "Atomic and Electronic Transport in Amorphous Ni-B Alloys", Thin Solid Films, 52 (1978) 89-96 (in French) (E)(see Phys. Abstr., 81(1978) 84325).

1190. J.FLECHON, F.A.KUHNAST, F.MACHIZAUD, B.AUGUIN and A.DEFRESNE, "Thermal Analysis of Amorphous Ni-P and Ni-B Alloys", J. Therm. Anal., 13 (1978) 241-52 (in French) (T)(see Chem. Abstr., 89(1978) 115999n).

1191. V.FLORESCU, "Amorphous Magnetic Materials", Lucr. Simp. Natl. Fiz. Solidului, eds. A.Nicula and O.Cozar, 1978, pp. 11-3 (in Romanian) (R)(see Chem. Abstr., 92(1980) 32902m).

1192. D.W.FORESTER, C.VITTORIA, D.C.WEBB and K.L.DAVIS, "Variable Delay Lines Using Magnetostrictive Metallic Glass Film Overlays", J. Appl. Phys., 49 (1978) 1794-6 (A).

1193. H.FRANKE, U.HEROLD, U.KÖSTER and M.ROSENBERG, "Mössbauer Spectroscopy and Electron Microscopy of Metastable Fe-B Alloys", J. Mag. Mag. Mater., 9 (1978) 214-7 (AB).

1194. H.FRANKE and M.ROSENBERG, "Mössbauer Spectroscopy of Amorphous Alloys (Metglasses)", J. Mag. Mag. Mater., 7 (1978) 168-74 (AQS, Mag).

1195. H.FRANKE, M.ROSENBERG, F.E.LUBORSKY and J.L.WALTER, "Mössbauer Spectroscopy of Ferromagnetic Metallic Glasses", IEEE Trans. Mag., MAG-14 (1978) 952-4 (Mag).

1196. R.L.FREED and J.B.VANDERSANDE, "A Study of the Crystallization of Two Non-Crystalline Copper-Zirconium Alloys", J. Non-Cryst. Solids, 27 (1978) 9-28 (AB).

1197. H.FUJIMORI, "Amorphous Magnetic Materials", Bull. Japan Inst. Metals, 17 (1978) 959-62 (in Japanese) (Mag, R).

1198. F.E.FUJITA, "Atomic Arrangement in the Amorphous Structure", Zairyo Kagaku, 15 (1978) 126-32 (in Japanese) (R)(see Chem. Abstr., 90(1979) 46660b).

1199. F.E.FUJITA, "Towards the Study of Structure of Amorphous Alloys", Metal Phys. Seminar, 3 (1978) 139-47 (in Japanese) (R)(see Chem. Abstr., 90(1979) 42350c).

1200. T.FUJIWARA, M.FUSE and Y.TANABE, "Density of States in Amorphous Metals", J. Phys. Soc. Japan, 44 (1978) 1397-8 (PP).

1201. K.FUKAMICHI, M.KIKUCHI, S.ARAKAWA, T.MASUMOTO, T.JAGIELINSKI, K.I.ARAI and N.TSUYA, "Magneto-Volume Effect in Invar-Type Amorphous Fe-B Alloys", Solid State Commun., 27 (1978) 405-7 (Mag).

1202. K.FUKAMICHI, M.KIKUCHI, H.HIROYOSHI and T.MASUMOTO, "Invar Effects in Amorphous Alloys", Solid State. Phys., 13 (1978) 322-32 (in Japanese) (R, Mag)(see Phys. Abstr., 82(1979) 20100).

1203. N.FUNAKOSHI, T.KANAMORI and T.MANABE, "Thermal Behaviour and Crystallization Kinetics of Amorphous Pd-Si Alloys", Japan. J. Appl. Phys., 17 (1978) 11-6 (T, AB).

1204. P.H.GASKELL, "A New Structural Model for Transition Metal-Metalloid Glasses", Nature, 276 (1978) 484-5 (SMA).

1205. W.GAUDIG and H.WARLIMONT, "The Structure of Short-Range Ordered β Cu-Al Alloys and a New Superlattice Phase", Acta Met., 26 (1978) 709-24 (AQS, MCS).

1206. R.GEMPERLE, L.KRAUS and J.SCHNEIDER, "Magnetization Reversal in Amorphous $(Fe_{1-x}Ni_x)_{80}P_{10}B_{10}$ Microwires", Czech. J. Phys., B28 (1978) 1138-45 (Mag)(see Met. Abstr., 12(1979) 33-1925).

1207. J.A.GERBER, D.J.MILLER and D.J.SELLMYER, "Magnetic Ordering in Amorphous Nd-Co, Gd-Co and Er-Co Alloys", J. Appl. Phys., 49 (1978) 1699-1701 (Mag).

1208. T.B.GHOSH, N.K.MISRA and G.B.MITRA, "An Apparatus for Liquisol Quenching", Indian J. Phys., A52 (1978) 299-304 (QT).

1209. M.R.J.GIBBS, J.P.PATTERSON, J.E.EVETTS and D.R.H.JONES, "Magnetic Properties of Homogeneously and Inhomogeneously Deformed Amorphous Alloys", Phys. Letters, A68 (1978) 461-2 (Mag).

1210. J.C.GIBELING and W.D.NIX, "A Study of the Creep Properties of a Ni-Fe Metallic Glass", Scripta Met., 12 (1978) 919-26 (M).

1211. B.C.GIESSEN, "The Determination and Representation of Metastable Phase Diagram Features and Other Thermal Characteristics of Metastable Alloys, Especially Amorphous Metals", in Applications of Phase Diagrams in Metallurgy and Ceramics, Part II (Nat. Bur. Standards, Washington, D.C., U.S.A., 1978) pp. 1161-85 (R)(see Phys. Abstr., 82(1979) 11568).

1212. B.C.GIESSEN and D.E.POLK, "Development and Initial Characterization of Amorphous Metals Rich in Tungsten and/or RE", UCRL Report 13989 (1978), 58pp (AS)(see Chem. Abstr., 91(1979) 197400p).

1213. J.J.GILMAN, "Structures of Ferrous Eutectic Liquids", Phil. Mag., B37 (1978) 577-84.

1214. E.GIRT, A.KURSUMOVIĆ and T.M.KOSANOVIC, "Anomalous Thermal Dilatation of Amorphous $Fe_{40}Ni_{40}B_{20}$", Fizika, 10, Suppl. 2 (1978) 267-71 (T)(see Chem. Abstr., 91(1979) 215286n).

1215. A.M.GLASS, K.NASSAU and T.J.NEGRAN, "Ionic Conductivity of Quenched Alkali Niobate and Tantalate Glasses", J. Appl. Phys., 49 (1978) 4808-11 (PP).

1216. A.A.GLAZER and R.I.TAGIROV, "Amorphous Magnetic Materials", Izv. Akad. Nauk SSSR, Ser. Fiz., 42 (1978) 1600-8 (in Russian); Bull. Acad. Sci. USSR, Phys. Ser., 42, No. 8 (1978) 27-34 (Mag) (see Phys. Abstr., 82(1979) 82839).

1217. B.GOLDING, J.E.GRAEBNER, A.B.KANE and J.L.BLACK, "Relaxation of Tunneling Systems by Conduction Electrons in a Metallic Glass", Phys. Rev. Letters, 41 (1978) 1487-91 (PP).

1218. B.GOLDING, J.E.GRAEBNER and W.H.HAEMMERLE, "Phonon Propagation in a Metallic Glass Below 1K", in Proc. Internat. Conf. on Lattice Dynamics, Paris, 1977 (Flammarion Sciences, Paris, 1978), pp. 348-50 (PP)(see Phys. Abstr., 81(1978) 71824).

1219. D.J.GOLDWASSER, E.P.OTOCKA and B.H.KEAR, "The Mechanical Behaviour of Polysulfone Matrix Composites Reinforced with Amorphous Metal Ribbons", Mater. Sci. & Engg., 34 (1978) 139-46 (M).

1220. U.GONSER, M.GHAFARI and H.-G.WAGNER, "Mössbauer Spectroscopy of Amorphous Metglas $Fe_{80}B_{20}$", J. Mag. Mag. Mater., 8 (1978) 175-7 (Mag).

1221. M.GOTO, H.TANGE and T.TOKUNAGA, "Magnetic Properties of the $(Fe_{1-x}Co_x)_{77}Si_{10}B_{13}$ Amorphous Alloys", Japan. J. Appl. Phys., 17 (1978) 1877-8 (Mag).

1222. T.GOTO, "Microstructure of Metallic Filament Produced by the Method of Melt Spinning", Nagoya Univ. HVEM Reports, No. 5 (1978) 12-7 (AQS).

1223. T.GOTO, "Tensile Strength of Stainless Steel Filament Produced by the Method of Glass-Coated Melt Spinning", Trans. Japan Inst. Metals, 19 (1978) 60-1 (M).

1224. T.GOTO, "A New Metastable Tetragonal Structure in Stainless Steel Filament Produced by Glass-Coated Melt Spinning", Trans. Japan Inst. Metals, 19 (1978) 605-10 (MCS).

1225. T.GOTO, "The Analysis of the Dynamics of the Melt Spinning of Metals Using the Spinnability of Glass", Sen-i Gakkaishi, 34 (1978) 237-44 (in Japanese) (PM).

1226. T.GOTO, "The Glass-Coated Melt Spinning of Gold Filaments", Gold Bulletin, 11 (1978) 124-5 (QT).

1227. C.D.GRAHAM, Jr. and T.EGAMI, "Magnetic Properties of Amorphous Alloys", Annual Rev. Mater. Sci., 8 (1978) 423-57 (R, Mag).

1228. W.A.GRANT, "Amorphous Metals and Ion Implantation", J. Vac. Sci. & Technol., 15 (1978) 1644-9 (R).

1229. B.GRÖGER and H.KRONMÜLLER, "Investigation of the Domain Structure, the Rayleigh Region and the Coercive Field of the Amorphous Ferromagnetic Alloy $Fe_{40}Ni_{40}P_{14}B_6$", J. Mag. Mag. Mater., 9 (1978) 203-7 (Mag).

1230. P.J.GRUNDY and J.M.MARSH, "Amorphous Thin Films of Stainless Steel", J. Mater. Sci., 13 (1978) 677-81 (Sp; AS).

1231. G.R.GRUZALSKI, J.A.GERBER and D.J.SELLMYER, "Magnetic Properties of Amorphous Transition Metal Alloys" in Transition Metals 1977, eds. M.J.G.Lee, J.M.Perz and E.Fawcett (Inst. Phys., London, 1978) pp. 445-8 (Mag)(see Chem. Abstr., 89 (1978) 121858g).

1232. G.R.GRUZALKSI, J.D.PATTERSON and D.J.SELLMYER, "Magnetic Ordering and Local Random Anisotropy in Dilute Magnetic Glasses", J. Appl. Phys., 49 (1978) 1696-8 (Mag).

1233. H.GUDMUNDSSON, H.U.ÅSTROM, D.NEW, K.V.RAO and H.S.CHEN, "Resistance Minima in Amorphous Mixed Magnetic System Fe-Mn-PBAl", J.Physique, 39 (1978) C6-943-5 (E).

1234. H.-J.GÜNTHERODT, M.MÜLLER, R.OBERLE, E.HAUSER, H.-U.KÜNZI, M.LIARD and R.MÜLLER, "Non-Crystalline Transition Metals", in Transition Metals 1977, eds. M.J.G.Lee, J.M.Perz and E.Fawcett (Inst. Phys., London, 1978) pp. 436-45 (E).

1235. P.HAASEN, "Metallic Glasses", Phys. Bl., 34 (1978) 573-84 (in German) (R)(see Phys. Abstr., 82(1979) 41962).

1236. T.D.HADNAGY, D.J.KRENITSKY, D.G.AST and C.Y.LI, "Load Relaxation Studies of a Metallic Glass", Scripta Met., 12 (1978) 45-8 (M).

1237. C.HARIGITAI and A.LOVAS, "Dependence of Magnetic Properties on the Composition in a Binary Metallic Glass System: Iron-Boron", Report KFKI-77-112, Hungarian Acad. Sci., Budapest (January 1978), 12pp (Mag)(see Phys. Abstr., 81(1978) 60314).

1238. R.HARRIS, M.SHALMON and M.J.ZUCKERMANN, "Negative Temperature Coefficient of Electrical Resistivity in Disordered Metallic Alloys", Phys. Rev., B18 (1978) 5906-11 (E).

1239. R.HARRIS and S.H.SUNG, "Metastable States in the Random Anisotropy Model for Amorphous Ferromagnets", J. Phys. F: Metal Phys., 8 (1978) L299-305 (Mag).

1240. R.HASEGAWA, "Reorientation Kinetics of the Magnetization in a Glassy Ferromagnet", J. Appl. Phys., 49 (1978) 5610-1 (Mag).

1241. R.HASEGAWA, M.C.NARASIMHAN and N.DE CRISTOFARO, "A High Permeability Fe-Ni Base Glassy Alloy Containing Mo", J. Appl. Phys., 49 (1978) 1712-4 (Mag).

1242. R.HASEGAWA and R.RAY, "Iron-Boron Metallic Glasses", J. Appl. Phys., 49 (1978) 4174-9 (Mag).

1243. R.HASEGAWA and L.E.TANNER, "Superconducting Transition Temperatures of Glassy and Partially Crystalline Be-Nb-Zr Alloys", J. Appl. Phys., 49 (1978) 1196-9 (S).

1244. K.HASHIMOTO, M.NAKA, K.ASAMI and T.MASUMOTO, "XPS and Electrochemical Studies of Effects of Metalloid Additives on Corrosion Behaviours of Amorphous Iron-Chromium Alloys", Corrosion Engg., 27 (1978) 279-83 (C).

1245. K.HASHIMOTO, M.NAKA, J.NOGUCHI, K.ASAMI and T.MASUMOTO, "Effect of Alloy Elements on Corrosion Resistance of Amorphous Iron-Base Alloys", in Passivity, eds. R.P.Frankenthal and J.Kruger (The Electrochem. Soc. Inc., Princeton, N.J., USA, 1978) pp. 156-69 (C).

1246. S.HATTA, T.EGAMI and C.D.GRAHAM Jr., "Amorphous Alloys with Improved Room Temperature Saturation Induction", IEEE Trans. Mag., MAG-14 (1978) 1013-5 (Mag).

1247. G.HAUSCH and E.TÖRÖK, "Elastic, Magnetoelastic and Thermal Properties of Some Ferromagnetic Metallic Glasses", Phys. Stat. Sol., (a) 50 (1978) 159-64 (T, M, Mag).

1248. E.HAUSER, R.J.ZIRKE, J.TAUC, J.J.HAUSER and S.R.NAGEL, "Optical Properties of Amorphous Metallic Alloys", Phys. Rev. Letters, 40 (1978) 1733-6 (Sp; PP).

1249. J.J.HAUSER, "Amorphous Nickel Films Getter Sputtered at 25 K", Phys. Rev., B17 (1978) 1908-12 (Sp; AQS, E).

1250. J.J.HAUSER and J.TAUC, "Electrical Properties of Binary Amorphous Alloys", Phys. Rev., B17 (1978) 3371-80 (E).

1251. J.J.HAUSER and J.V.WASZCZAK, "Magnetic Susceptibility of an Amorphous Spin Glass : Au-Si-Mn", Phys. Rev., B18 (1978) 6206-12 (Mag).

1252. T.M.HAYES, "EXAFS Studies of Disordered Solids", J. Non-Cryst. Solids, 31 (1978) 57-79 (AQS, R).

1253. T.M.HAYES, J.W.ALLEN, J.TAUC, B.C.GIESSEN and J.J.HAUSER, "Short-Range Order in Metallic Glasses", Phys. Rev. Letters, 40 (1978) 1282-5 (AQS).

1254. N.HEIMAN, R.D.HEMPSTEAD and N.KAZAMA, "Low Coercivity Amorphous Magnetic Alloy Films", J. Appl. Phys., 49 (1978) 5663-7 (Mag).

1255. N.HEIMAN and N.S.KAZAMA, "Magnetization of Amorphous RE_xCu_{1-x} Alloys (RE=Gd, Tb, Dy, Ho)", J. Appl. Phys., 49 (1978) 1686-8 (Mag).

1256. N.HEIMAN and N.S.KAZAMA, "Concentration Dependence of the Co Moment in Amorphous Alloys of Co with Y, La and Zr", Phys. Rev., B17 (1978) 2215-20 (Mag).

1257. N.HEIMAN, N.S.KAZAMA, D.F.KYSER and V.J.MINKIEWICZ, "Effect of Substrate Bias and Annealing on the Properties of Amorphous Alloy Films of Gd-Co, Gd-Fe and Gd-Co-X (X=Mo, Cu, Au)", J. Appl. Phys., 49 (1978) 366-75 (Mag).

1258. M.v.HEIMENDAHL and R.OPPOLZER, "Direct Measurement of Decomposition on Crystallization of Metglas 2826A by X-Ray Microanalysis in STEM", Scripta Met., 12 (1978) 1087-90 (AB).

1259. B.HEINRICH, T.HAAG and J.GEIGER, "A Metallic Glass Tip: A Promising Field Electron Emission Source", J. Phys. D: Appl. Phys., 11 (1978) 2439-42 (A).

1260. U.HEROLD and U.KÖSTER, "Metastable Phases in Rapidly Solidified Iron-Boron Alloys", Z. Metallkde., 69 (1978) 326-32 (in German) (MCS, AB, AS).

1261. H.-R.HILZINGER, A.MAGER and H.WARLIMONT, "Amorphous Ferromagnetic Materials - Magnetic Fundamentals, Properties and Applications", J. Mag. Mag. Mater., 9 (1978) 191-9 (in German) (R, Mag).

1262. W.A.HINES, L.T.KABACOFF, R.HASEGAWA and P.DUWEZ, "NMR Knight Shifts and line Widths in the Ni-Pd-P and Ni-Pt-P Metallic Glasses : Composition and Temperature Dependences", J. Appl. Phys., 49 (1978) 1724-6 (Mag).

1263. E.HIROTA, "Amorphous Magnetic Alloy Materials and Their Applications", Zairyo Kagaku, 15 (1978) 152-5 (in Japanese) (R)(see Chem. Abstr., 90(1979) 65567t).

1264. H.HIROYOSHI, K.FUKAMICHI, M.KIKUCHI, A.HOSHI and T.MASUMOTO, "Magnetic Moment and High-Field Susceptibility of Amorphous FeB Invar-Type Alloys", Phys. Letters, A65 (1978) 163-5 (Mag).

1265. M.R.HOARE, "Packing Models and Structural Specificity", J. Non-Cryst. Solids, 31 (1978) 157-79 (SMA).

1266. O.HORIE, Y.SYONO Y.NAKAGAWA, A.ITO, K.OKAMURA and S.YAJIMA, "Mössbauer Study of Amorphous $BaO-Fe_2O_3-B_2O_3$ System", Solid State Commun., 25 (1978) 423-6 (AQS).

1267. E.HORNBOGEN, "Structure and Thermal Stability of Vapour-Deposited Fe-Cu Alloys", J. Mater. Sci., 13 (1978) 666-70 (VD; AB, AQS).

1268. S.K.HSIUNG and R.WANG, "Thermoelectric Properties of Splat-Cooled Amorphous $In_{20}Te_{80}$, $Ga_{20}Te_{80}$ and $Ge_{15}Te_{85}$", J. Appl. Phys., 49 (1978) 280-4 (T).

1269. S.HUDSON and H.C.ANDERSON, "The Glass Transition of Atomic Glasses", J. Chem. Phys., 69 (1978) 2323-31 (T).

1270. O.HUNDERI, "Metallic Glass - A New Material Form. Fascinating Application Perspectives", Tek. Ukebl., 125 (1978) 25-6 (in Norwegian) (R)(see Chem Abstr., 89(1978) 133853q).

1271. O.HUNDERI, "Metallic Glasses - Manufacture, Structure and Stability. I", Fra. Fys. Verden (Norway), 40(1978) 44-7 (in Norwegian) (QT, AS)(see Phys. Abstr., 82(1979) 14687).

1272. O.HUNDERI, "Metallic Glass - Properties. II", Fra. Fys. Verden (Norway), 40 No. 4 (1978) 89-92 (in Norwegian)(M, Mag, PP) (see Phys. Abstr., 82(1979) 32909).

1273. Gh. ILONCA, "Magnetic Properties of Amorphous $Mn_xPd_{82-x}Ge_{18}$ Alloys", Czech. J. Phys., B28 (1978) 1132-7 (Mag)(see Met. Abstr., 12(1979) 33-1924).

1274. Gh. ILONCA, "Magnetic Properties of Amorphous $Cr_xPd_{82-x}Ge$ Alloys", Canad. J. Phys., 56 (1978) 1084-7 (mag).

1275. Gh. ILONCA, "Electrical Resistivity and Magnetoresistivity of Amorphous $Cr_xPd_{82-x}Ge_{18}$ Alloys", Phys. Stat. Sol., (b) 88 (1978) 37-41 (E, Mag).

1276. A.INOUE, S.ARAKAWA and T.MASUMOTO, "In-Situ Transformation of Cementite to M_7C_3 and Internal Defects of M_7C_3 in High Carbon-Chromium Steel by Tempering", Trans. Japan Inst. Metals, 19 (1978) 11-7 (AB).

1277. A.INOUE and T.MASUMOTO, "Transmission Electron Microscopic Study on In-Situ Carbide Reactions during Tempering of High-Carbon High-Chromium Steels Rapidly Quenched from Melts", J. Japan Inst. Metals, 42 (1978) 787-95 (in Japanese)(AB)(see Met. Abstr., 12(1979) 11-0477).

1278. A.INOUE, T.MASUMOTO, S.ARAKAWA and T.IWADACHI, "Amorphous High-Carbon Alloy Steels Rapidly Quenched from Melts", Trans. Japan Inst. Metals, 19 (1978) 303-4 (AS).

1279. A.INOUE, T.MASUMOTO, M.KIKUCHI and T.MINEMURA, "Compositional Effect on Crystallization of (Fe, Ni, Co)-Si-B Amorphous Alloys", J. Japan Inst. Metals, 42 (1978) 294-303 (in Japanese) (AB) (see Met. Abstr., 11(1978) 12-1174).

1280. A.INOUE, T.MASUMOTO and H.M.KIMURA, "Embrittlement of Fe, Ni Co Amorphous Alloys Containing Metalloid Elements by Crystallization", J. Japan Inst. Metals, 42 (1978) 303-9 (in Japanese) (M)(see Met. Abstr., 11(1978) 31-3092).

1281. J.IVKOV, Ž.MAROHNIĆ, E.BABIĆ, D.PIVAC and J.ŠUBIĆ, "Magnetic Contribution to Resistivity of Amorphous $Fe_xNi_{80-x}P_{14}B_6$ Alloys", Fizika, 10, Suppl. 2 (1978) 282-5 (E)(see Phys. Abstr., 82(1979) 56182).

1282. H.IWASAKI and T.MASUMOTO, "Effect of High Pressure on the Crystallization of an Amorphous $Pd_{80}Si_{20}$ Alloy", J. Mater. Sci., 13 (1978) 2171-6 (AB).

1283. T.JAGIELINSKI, K.I.ARAI, N.TSUYA and K.FUKAMICHI, "Magnetoelastic Effects in Invar-Type Amorphous Fe-B Alloys", Phys. Stat. Sol., (a) 50 (1978) K25-8 (Mag).

1284. S.JANOS, A.FEHER, M.REIFFERS, M.KONC and O.DUSA, "Electrical and Magnetic Properties of Amorphous NiCoP Alloys", in Transition Metals 1977, eds. M.J.G.Lee, J.M.Perz and E.Fawcett (Inst. Phys., London, 1978), pp. 448-52 (E, Mag)(see Chem. Abstr., 89(1978) 121859h).

1285. C.JANTZEN, D.SCHWAHN, J.SCHELTEN and H.HERMAN, "Phase Decomposition in Al_2O_3-SiO_2 Glasses", J. Appl. Cryst., 11 (1978) 614-5 (AB).

1286. W.L.JOHNSON and S.J.POON, "Superconductivity of Amorphous $(Mo_{1-x}Ru_x)_{80}P_{20}$ Alloys", J. Physique, 39 (1978) C6-451-3 (S).

1287. W.L.JOHNSON, S.J.POON, J.DURAND and P.DUWEZ, "Superconductivity, Magnetic Susceptibility and Electronic Properties of $(Mo_{1-x}Ru_x)_{80}P_{20}$ Alloys Obtained by Liquid Quenching", Phys. Rev., B18 (1978) 206-17 (S, Mag).

1288. W.L.JOHNSON, C.C.TSUEI and P.CHAUDHARI, "Paraconductivity of Three-Dimensional Amorphous Superconductors - Evidence for a Short Wavelength Cutoff in the Fluctuation Spectrum", Phys. Rev., B17 (1978) 2884-91 (S).

1289. H.JONES, "Developements in Aluminum Alloys by Solidification at Higher Cooling Rates", Aluminium, 54 (1978) 274-81 (R).

1290. M.KACZOROWSKI, J.KOZUBOROSKI, B.DABROWSKI and H.MATYJA, "Crystallization of Amorphous Te-20 at.% Pb Alloy Obtained by Rapid Cooling from the Liquid State", J. Mater. Sci., 13 (1978) 407-14 (AS, AB).

1291. M.KACZOROWSKI, J.KOZUBOROSKI, B.DABROWSKI and H.MATYJA, "Crystallization of Amorphous Te-20 at.% Sn Alloy", J. Mater. Sci., 13 (1978) 1105-13 (AB).

1292. G.KAECH and W.EPPRECHT, "New Forms of Metastable Precipitates and Amorphous Phases in a Laser Irradiated Al-2.12 at.% Cu Alloy", Scripta Met., 12 (1978) 493-6 (in French) (AS, MCS) (see Chem. Abstr., 90(1979) 42478a).

1293. T.KANEYOSHI, "Comment on the Possibility of a New Low Temperature Phase in Amorphous Magnetic Metals", J. Phys. Soc. Japan, 45 (1978) 1411 only (MCS).

1294. T.KANEYOSHI, "On the Paramagentic Curie Temperature in Amorphous Ferromagnetic Metals", J. Phys. Soc. Japan, 45 (1978) 94-8 (Mag).

1295. K.KAWANO, H.INO and S.NISHIKAWA, "A Study of Liquid-Quenched La-Fe Alloys", Scripta Met., 12 (1978) 333-5 (AQS).

1296. N.S.KAZAMA and N.HEIMAN, "Magnetic Properties of Amorphous FeC Thin Films", J. Appl. Phys., 49 (1978) 1706-8 (Mag).

1297. T.KEMÈNY, B.FOGARASSY and E.TOTH-KADAR, "Crystallization Kinetics of Metallic Glasses", Report KFKI-78-10, Hungarian Acad. Sci., Budapest (February 1978), 8pp (AB)(see Phys. Abstr., 81(1978) 91257).

1298. M.KIKUCHI, K.FUKAMICHI, T.MASUMOTO, T.JAGIELINSKI, K.I.ARAI and N.TSUYA, "Giant ΔE Effect and Elinvar Characteristics in Amorphous Fe-B Binary Alloys", Phys. Stat. Sol., (a) 48 (1978) 175-81 (M).

1299. R.J.KOBLISKA, J.A.ABOAF, A.GANGULLEE, J.J.CUOMO and E.KLOKHOLM, "Amorphous Ferromagnetic Thin Films", Appl. Phys. Letters, 33 (1978) 473-5 (Mag).

1300. O.KOHMOTO, N.YAMAGUCHI, K.OHYA and H.FUJISHIMA, "Change of the Magnetic Property in Some Amorphous Alloys by Low Temperature Annealing", Japan. J. Appl. Phys., 17 (1978) 257-8 (Mag).

1301. O.KOHMOTO, N.YAMAGUCHI, K.OHYA, H.FUJISHIMA and T.OJIMA, "Effects of Annealing on Magnetostriction and Permeability of Zero Magnetostrictive Amorphous Alloys", IEEE Trans. Mag., MAG-14 (1978) 949-51 (Mag).

1302. M.KONC, J.DANIEL-SZABO and O.DUSA, "Contribution to the Study of the Creep Effect in Thin Amorphous Electrodeposited CoP and NiCoP Films", Acta Phys. Slovaca, 28 (1978) 77-80 (Mag) (see Phys. Abstr., 81(1978) 60322).

1303. M.KONC, J.DANIEL-SZABO, A.KOSTURIAK and O.DUSA, "Preparation and Some Basic Physical Properties of the Elctrodepostied NiCoP Films", Acta Phys. Slovaca, 28 (1978) 57-64 (Mag, AS) (see Phys. Abstr., 81(1978) 60587).

1304. D.KORYTÁR and P.MRAFKO, "Similarity and Short-Range Order Parameter for Liquids and Amorphous Solids", Acta Phys. Slovaca, 28, No. 2 (1978) 108-19 (SMA)(see Met. Abstr., 11(1978) 15-1565).

1305. U.KÖSTER and U.HEROLD, "Crystallization of Amorphous $Fe_{80}B_{20}$", Scripta Met., 12 (1978) 75-7 (AB).

1306. J.KREMPASKÝ, "A Theory of Electrical Conductivity in Metallic Glasses", Czech. J. Phys., B28 (1978) 653-62 (E)(see Phys. Abstr., 81(1978) 60102).

1307. U.KREY, "Spin Wave Excitations and the Magnetization Curve of Amorphous Ferromagnets", Z. Physik, B31 (1978) 247-55 (Mag).

1308. H.KREYE, M.HAMMERSCHMIDT and G.REINERS, "Ultrasonic Welding of Metallic Alloy Glass to Copper", Scripta Met., 12 (1978) 1059-61 (F).

1309. E.S.KUCHERENKO, "Conditions for Dendritic Crystallization of Bi-Sb Alloy Accompanying Rapid Cooling", Izv. Akad. Nauk SSSR Met., No. 4 (1978) 115-9 (in Russian) (PM)(see Phys. Abstr., 82(1979) 34020).

1310. T.KUDO, T.MIZOGUCHI, N.WATANABE, N.NIMURA, M.MISAWA and K.SUZUKI, "Partial Structure Factor of Amorphous $Cu_{0.57}Zr_{0.43}$ Alloy Determined by ToF Pulsed Neutron Diffraction", J. Phys. Soc. Japan, 45 (1978) 1773-4 (AQS).

1311. U.KULMANN and G.DIETZ, "Ferromagnetic Resonance in Amorphous Co-P Alloys", J. Mag. Mag. Mater., 9 (1978) 211-3 (ED; Mag).

1312. E.LAINE, I.LAHTEENMAKI and I.LEHTORANTA, "X-Ray Study of Splat-Quenched Sn-Bi and Cu-Sb Alloys by Energy Dispersive Diffraction", J. Mater. Sci., 13 (1978) 108-12 (AQS).

1313. M.LASOCKA, "Thermal Evidence of Overlapping Effects of Glass Transition and Crystallization, Derived from two different Glassy Phases in Phase-Separated System $Te_{80}Ge_{12.5}Pb_{7.5}$", 13 (1978) 2055-9 (T).

1314. M.LASOCKA, "Thermal Evidence of Structural Phase Transformation Concurrent with Glass Crystallization Effects in Te-Rich Ternary Systems Te-Ge-A (A=Sn, Pb)", J. Mater. Sci., 13 (1978) 2514-7 (T).

1315. M.LASOCKA, "Formal $\overline{\Delta H}_v$ versus $1/\overline{M}$ curves used for Prediction of the Tendency of Changes in Glass Transition Temperature Independence on Concentration in Te-Rich Ternary Chalcogenides", Phys. Letters, A65 (1978) 256-8 (T).

1316. J.LATUSZKIEWICZ, M.STREZESZEWSKA, A.CALKA, P.G.ZIELINSKI and H.MATYJA, "Thermal Stability of Amorphous $Cu_{60}Me_{40}$ Alloys (Me=Cd, Zr)", Metals Tech., 5 (1978) 329-32 (T).

1317. H.J.LEAMY and A.G.DIRKS, "Microstructure and Magnetism in Amorphous Rare-Earth-Transition Metal Thin Films. I. Microstructure", J. Appl. Phys., 49 (1978) 3430-8 (VD; Mag).

1318. F.J.LITTERST, A.OGRODNIK and G.M.KALVIUS, "Magnetic Hyperfine Data of Non-Crystalline Iron from Mössbauer Spectroscopy", Hyperfine Interactions, 4 (1978) 879-85 (AQS)(see Phys. Abstr., 81 (1978) 39477).

1319. V.A.LOBODYUK and V.K.TKACHUK, "The Structure of a Copper-25 wt.% Sn Alloy on Splat Cooling", Izv. Akad. Nauk SSSR Met., No. 2 (1978) 105-9 (in Russian); Russian Metall., No. 2 (1978) 88-91 (AQS)(see Met. Abstr., 11(1978) 12-1294).

1320. F.E.LUBORSKY, "Magnetic Properties of Amorphous Alloys", J. Mag. Mag. Mater., 7 (1978) 143-9 (Mag).

1321. F.E.LUBORSKY, "Applications of Amorphous Alloys", IEEE Trans. Mag., MAG-14 (1978) 1008-12 (A, Mag).

1322. F.E.LUBORSKY, "Amorphous Ferromagnets", in Ferromagnetic Materials, ed. E.P.Wohlfarth (North-Holland Publishing Co., Amsterdam, 1978) (Mag).

1323. F.E.LUBORSKY, "Amorphous Magnetic Materials", in Kirk-Othmer Encyclo. Chem. Tech. (Third edition), eds. M.Grayson, D.Eckroth (Wiley, New York, 1978), Vol. 2, pp. 537-69 (R) (see Chem. Abstr., 89(1978) 15694r).

1324. F.E.LUBORSKY, J.J.BECKER, P.G.FRISCHMANN and L.A.JOHNSON, "Potential of Amorphous Alloys for Application in Magnetic Devices", J. Appl. Phys., 49 (1978) 1769-74 (A, Mag).

1325. F.E.LUBORSKY, P.G.FRISCHMANN and L.A.JOHNSON, "The Role of Amorphous Materials in the Magnetic Industry", J. Mag. Mag. Mater., 8 (1978) 318-29 (R, Mag).

1326. F.E.LUBORSKY, G.KOTE and S.ARAJS, "Magnetic and Electrical Properties of Amorphous $(Fe_xNi_{1-x})_{80}B_{20}$ Alloys at Elevated Temperature", Bull. Amer. Phys. Soc., 23 (1978) 332 (E, Mag).

1327. F.E.LUBORSKY and H.H.LIEBERMANN, "Crystallization Kinetics of Fe-B Amorphous Alloys", Appl. Phys. Letters, 33 (1978) 233-4 (AB).

1328. F.E.LUBORSKY and J.L.WALTER, "Stress Relaxation In Amorphous Alloys", Mater. Sci. & Engg., 35 (1978) 255-61 (M).

1329. A.B.LYSENKO and A.A.YAKUNIN, "Formation of Metastable and Non-Crystalline Phase in Rapidly Cooled La-Ag Alloys", Izv. Vyssh. Uchebn. Zaved., Tsvetn. Metall., No. 1 (1978) 96-101 (in Russian) (MCS, AS)(see Chem Abstr., 88(1978) 159246y).

1330. K.MAEDA and S.TAKEUCHI, "Deformation Mechanism of Amorphous Metals", Oyo Buturi, 47 (1978) 878-81 (in Japanese) (M)(see Phys. Abstr., 82(1979) 42136).

1331. K.MAEDA and S.TAKEUCHI, "A Simple Computer Modelling of Metallic Amorphous Structure", J. Phys. F: Metal Phys., 8 (1978) L283-8 (SMA).

1332. K.MAEDA and S.TAKEUCHI, "Computer Simulation of Deformation in Two-Dimensional Amorphous Structures", Phys. Stat. Sol., (a) 49 (1978) 685-96 (M).

1333. A.J.MAELAND, "Comparison of Hydrogen Absorption in Glassy and Crystalline Structures", Hydrides for Energy Storage, eds. A.F.Andresen and A.J.Maeland (Pergamon Press, Oxford, U.K., 1978) pp. 447-62 (A)(see Met. Abstr., 12(1979) 34-0348).

1334. R.MALMHÄLL, G.BÄCKSTRÖM, S.M.BHAGAT and K.V.RAO, "Hall Resistivity of a Magnetic Amorphous Alloy - Effects of Thermal Cycling and Annealing", J. Non-Cryst. Solids, 28 (1978) 159-76 (Mag, AB).

1335. R.MALMHÄLL. G.BÄCKSTRÖM. K.V.RAO. S.M.BHAGAT, M.MEICHLE and M.B.SALAMON, "Metglas 2826B - Transport, Magnetic and Thermal Properties", J. Appl. Phys., 49 (1978) 1727-9 (T, Mag).

1336. Ph. MANGIN and G.MARCHAL, "Electrical Resistivity and Ziman's Formula in the Amorphous Alloy Fe_xSn_{1-x}", Phys. Letters, A68 (1978) 466-8 (VD; E).

1337. Ph. MANGIN and G.MARCHAL, "Structural and Magnetic Properties in Fe_xSi_{1-x} Amorphous Alloys", J. Appl. Phys., 49 (1978) 1709-11 (VD; Mag).

1338. J.S.MANNING and C.V.BRISCOE, "Stability, Superconducting Transition Temperature and Residual Resistivity of Amorphous Films of $La_{100-x}Au_x$ Alloys", Phys. Rev., B18 (1978) 1177-82 (S, T, E).

1339. G.MARCHAL, B.RODMACQ, Ph.MANGIN, M.PIECUCH and Chr.JANOT, "Composition and Temperature Ranges for Amorphous Fe_xSn_{1-x} Alloy Stability", Mater. Sci. & Engg., 36 (1978) 11-5 (VD; AS).

1340. R.E.MARINGER and C.E.MOBLEY, "The Melt Extraction of Metallic Filament and Staple Fiber", AIChE Symp. Series No. 180 - Spinning Wire from Molten Metal, 74 (1978) 16-9 (QT).

1341. R.E.MARINGER, C.E.MOBLEY and E.W.COLLINGS, "Preparation and Properties of Compacts of Melt Extract Staple Fibers of Ti-6Al-4V Alloy", AIChE Symp. Series No. 180 - Spinning Wire from Molten Metal, 74 (1978) 111-6 (QT).

1342. Ž.MAROHNIĆ, A.HAMZIĆ, E.BABIĆ and J.IVKOV, "Galvanomagnetic Properties of Amorphous Ferromagnets", Fizika, 10, Suppl. 2 (1978) 286-9 (Mag)(see Phys. Abstr., 82(1979) 56196).

1343. Ž.MAROHNIĆ, J.IVKOV and E.BABIĆ, "Hall Effect in Amorphous $Fe_xNi_{80-x}B_{20}$ and $Fe_xNi_{80-x}P_{14}B_6$ Alloys", Electrotehnika (Zagreb), No. 5-6 (1978) 7 415-9 (in Croatian) (Mag) (see Phys. Abstr., 82(1979) 11102).

1344. O.MASSENET and H.DAVER, "Low Temperature Conversion Electron Mössbauer Spectroscopy on Fe_xGe_{1-x} Amorphous Thin Films", Solid State Commun., 25 (1978) 917-20 (VD; Mag).

1345. T.MASUMOTO, "Preparation Techniques for Amorphous Materials", Oyo Buturi, 47 (1978) 835-8 (in Japanese) (QT)(see Phys. Abstr., 82(1979) 43169).

1346. T.MASUMOTO, "Future Applications of Amorphous Magnetic Alloys", Machine Design, 5 (1978) 52-6 (in Japanese) (R).

1347. T.MASUMOTO, "Amorphous Metals", Science, 8 (1978) 62-72 (in Japanese) (R).

1348. T.MASUMOTO, "Amorphous Soft Magnetic Alloys", Tech. & Market Report, (1978) 87-96 (in Japanese) (R).

1349. T.MASUMOTO, "Metallurgical Problems on Amorphous Magnetic Alloys", in Proc. 23rd Conf. Magnetism and Magnetic Materials, (1978) 57-63 (in Japanese) (R).

1350. T.MASUMOTO, "Prospect of Amorphous Alloys from Engineering Point of View", Oyo Buturi, 47 (1978) 862-6 (in Japanese) (R)(see Chem. Abstr., 90(1979) 125707a).

1351. T.MASUMOTO, "Amorphous Magnetic Alloys", in Proc. Conf. Fundamental and Applications of Magnetism (J. Inst. Appl. Mag., 1978) pp. 41-54 (R).

1352. T.MASUMOTO and K.HASHIMOTO, "Chemical Properties of Amorphous Metals", Annual Rev. Mater. Sci., 8 (1978) 215-33 (C, R).

1353. J.R.MATEY and A.C.ANDERSON, "Phonon Propagation in Glassy Metals", Phys. Rev., B17 (1978) 5029-32 (PP).

1354. T.R.McGUIRE, T.MIZOGUCHI, R.J.GAMBINO and S.KIRKPATRICK, "Magnetic Phase Diagram of the Gd-Al and Gd-Cu Amorphous Alloy Systems", J. Appl. Phys., 49 (1978) 1689-90 (Mag).

1355. R.MEHRABIAN, S.KOU, S.C.HSU and A.MUNITZ, "Laser Surface Melting and Subsequent Solidification", AIP Conf. Proc., 50 (1978) 129-48 (QT, AQS) (R)(see Phys. Abstr., 82(1979) 69916).

1356. L.V.MEISEL and P.J.COTE, "Critical Test of the Diffraction Model in Amorphous and Disordered Metals", Phys. Rev., B17 (1978) 4652-9 (ED; T).

1357. A.G.MIKOLAICHUK, A.N.KOGUT, V.I.KOVAL'CHUK and P.I.SHEVCHUK, "A Study of the X-Ray Spectra of Amorphous and Crystalline Germanium Telluride", Ukr. Fiz. Zhur., 23 (1978) 1393-5(AQS) (see Met. Abstr., 12(1979) 12-0724).

1358. O.MILAT and D.KUNSTELJ, "Crystllization of Amorphous Zr-Fe Alloy Quenched by the Levitation Method", Fizika, 10, Suppl.2 (1978) 305-10 (AB)(see Chem. Abstr., 92(1980) 9891y).

1359. I.S.MIROSHNICHENKO and V.F.BASHEV, "Structure and Properties of Rapidly Solidified Fe-B, Fe-P and Co-B Alloys", Metall. Gornorudn. Prom- st., No. 4 (1978) 40-2 (in Russian) (TSSE, AS)(see Chem. Abstr., 90(1979) 190911b).

1360. M.A.MITCHELL, A.E.CLARK, H.T.SAVAGE and R.J.ABBUNDI, "ΔE Effect and Magnetomechanical Coupling Factor in $Fe_{80}B_{20}$ and $Fe_{78}Si_{10}B_{12}$ Glassy Ribbons", IEEE Trans. Mag., MAG-14 (1978) 1169-71 (M, Mag).

1361. M.MITERA, M.NAKA, T.MASUMOTO, N.KAZAMA and K.WATANABE, "Effect of Metalloids on the Magnetic Properties of Iron-Based Amorphous Alloys", Phys. Stat. Sol., (a) 49 (1978) K163-6 (Mag).

1362. T.MIYAZAKI and M.TAKAHASHI, "Magnetic Annealing Effect of Amorphous $(Fe_{1-x}Co)_{77}Si_{10}B_{13}$ Alloys", Japan. J. Appl. Phys., 17 (1978) 1755-63 (Mag).

1363. U.MIZUTANI, K.T.HARTWIG, T.B.MASSALSKI and R.W.HOPPER, "Low Temperature Specific Heats of Glassy $Pd_{1-x-y}Si_xCu_y$ Alloys", Phys. Rev. Letters, 41 (1978) 661-4 (T).

1364. K.MOHRI and S.KOREKODA, "New Force Transducers Using Amorphous Ribbon Cores", IEEE Trans. Mag., MAG-14 (1978) 1071-5 (A).

1365. J.B.MONTEIL, L.PADEL and J.C.BERNIER, "Structure and Crystallization Mechanism of the Products Obtained by Splat Quenching in the $BaO-Fe_2O_3$ and $SrO-Fe_2O_3$ Systems", J. Solid State Chem., 25 (1978) 1-8 (in French) (AQS, AB)(see Phys. Abstr., 81(1978) 51426).

1366. H.A.MOOK, "Magnetic Excitations in Amorphous Ferromagnets", J. Appl. Phys., 49 (1978) 1665-70 (Mag).

1367. P.MRAFKO, "Homogeneous and Isotropic Hard Sphere Model of Non-Crystalline Solids", Acta Phys. Slovaca, 28, No. 2 (1978) 120-4 (SMA)(see Met. Abstr., 11(1978) 15-1566).

1368. H.R.MÜLLER, "Amorphous Magnetic Materials", Wiss. Fortschr., 28, No. 2 (1978) 69-73 (in German) (R)(see Chem. Abstr., 89 (1978) 35407c).

1369. M.MÜLLER, H.BECK and H.-J.GÜNTHERODT, "Magnetic Properties of Liquid Pd, Si and Pd-Si Alloys", Phys. Rev. Letters, 41 (1978) 983-7 (Mag).

1370. S.R.NAGEL, "The Polk Model and the Structure of Metallic Glasses", Scripta Met., 12 (1978) 43-4 (SMA).

1371. S.R.NAGEL, "Thermoelectric Power and Resistivity in a Metallic Glass", Phys. Rev. Letters, 41 (1978) 990-3 (T, E).

1372. S.R.NAGEL, J.VASSILIOU, P.M.HORN and B.C.GIESSEN, "Temperature Dependence of the Resistivity of Nb-Ni Glasses", Phys. Rev., B17 (1978) 462-7 (E).

1373. M.NAKA, K.HASHIMOTO and T.MASUMOTO, "Effect of Metalloidal Elements on Corrosion Resistance of Amorphous Iron-Chromium Alloys", J. Non-Cryst. Solids, 28 (1978) 403-13 (C).

1374. M.NAKA, K.HASHIMOTO and T.MASUMOTO, "High Corrosion Resistance of Amorphous Fe-Mo and Fe-W Alloys in HCl", J. Non-Crys. Solids, 29 (1978) 61-5 (C).

1375. M.NAKA, K.HASHIMOTO and T.MASUMOTO, "Corrosion Behaviour of Amorphous and Crystalline $Cu_{50}Ti_{50}$ and $Cu_{50}Zr_{50}$ Alloys", J. Non-Cryst. Solids, 30 (1978) 29-36 (C).

1376. K.NARITA, J.YAMASAKI and H.FUKUNAGA, "Composition Dependence of Magnetic Moment and Curie Temperature in Amorphous $(Fe_x$-$Co_{1-x})_{100-(y+x)}Si_yB_x$ Ribbons", IEEE Trans. Mag., MAG-14 (1978) 1016-8 (Mag).

1377. H.NEUHÄUSER, "Rate of Shear Band Formation in Metallic Glasses", Scripta Met., 12 (1978) 471-4 (M).

1378. A.K.NIGAM and A.K.MAJUMDAR, "Ferromagnetic Anisotropy of Resistivity in Metglas", Physica, B+C95 (1978) 385-90 (Mag).

1379. T.NINOMIYA and H.KOIZUMI, "Dislocation Model of Amorphous Structure", Oyo Buturi, 47 (1978) 882-6 (in Japanese) (SMA) (see Phys. Abstr., 82(1979) 42065).

1380. R.C.O'HANDLEY, "Magnetostriction of Transition Metal-Metalloid Glasses: Temperature Dependence", Phys. Rev., B18 (1978) 930-8 (Mag).

1381. R.C.O'HANDLEY, "Spontaneous Hall Effect and Resistivity of Fe-Co-Ni-Base Glasses", Phys. Rev., B18 (1978) 2577-82 (Mag).

1382. R.C.O'HANDLEY and L.BERGER, "Evidence Supporting the Split-Band Model from Magnetostrictions of Transition Metal/Metalloid Glasses", in Transition Metals 1977 eds. M.J.G.Lee, J.M.Perz and E.Fawcett (Inst. Phys., London, 1978) pp. 477-81 (Mag).

1383. R.C.O'HANDLEY and D.S.BOUDREAUX, "Magnetic Properties of Transition Metal-Metalloid Glasses : A Charge Transfer Model", Phys. Stat. Sol., (a) 45 (1978) 607-15 (Mag).

1384. R.C.O'HANDLEY and C.P.CHOU, "Magnetoelastic Effects in Metallic Glasses", J. Appl. Phys., 49 (1978) 1659-64 (Mag).

1385. S.N.OJHA and K.CHATTOPADHYAY, "Electron Microscopic Investigation of Liquisol-Quenched Monotectic Aluminium-Indium Alloy", Trans. Indian Inst. Metals, 31 (1978) 208-10 (TSSE, MCS).

1386. D.G.ONN, T.H.ANTONIUK, T.A.DONNELLY, W.D.JOHNSON, T.EGAMI, J.T.PRATER and J.DURAND, "Specific Heat and Susceptibility of Ni-Based Amorphous Alloys with Dilute Fe", J. Appl. Phys., 49 (1978) 1730-2 (T, Mag).

1387. R.S.OSWALD, M.RON and M.OHRING, "Conversion Electron Mössbauer Studies of Amorphous Fe-Si Thin Films", J. Vac. Sci. & Tech., 15 (1978) 637 (VD; AQS) (Abstract only).

1388. C.A.PAMPILLO and D.E.POLK, "Annealing Embrittlement in an Iron-Nickel-Based Metallic Glass", Mater. Sci. & Engg., 33 (1978) 275-80 (AB).

1389. G.W.F.PARODE, E.BUTLER and D.GELDER, "Rapid Quenching by the Taylor Wire Technique", J. Mater. Sci., 13 (1978) 786-90 (AS).

1390. J.D.PATTERSON, G.R.GRUZALSKI and D.J.SELIMYER, "Effect of Random Anisotropy on Magnetic Properties of Amorphous Systems", Phys. Rev., B18 (1978) 1377-90 (Mag).

1391. J.P.PATTERSON and D.R.H.JONES, "Mouliding of a Metallic Glass", Mater. Res.Bull., 13 (1978) 583-5 (F).

1392. D.PIVAC, Z.VUČIĆ, Ž.MAROHNIĆ and E.BABIĆ, "Thermoelectric Power of Amorphous Ferromagnets near Curie Temperature", Fizika, 10 Suppl. 2 (1978) 277-81 (T)(see Phys. Abstr., 82 (1979) 56200).

1393. J.M.POATE, "Metastable Alloy Formation", J. Vac. Sci. & Tech., 15 (1978) 1636-43 (R).

1394. D.E.POLK, A.CALKA and B.C.GIESSEN, "Preparation and Thermal and Mechanical Properties of New Titanium-Rich Metallic Glasses", Acta Met., 26 (1978) 1097-1103 (AS, T, M).

1395. S.J.POON, "The Range of Indirect Exchange Interaction in an Amorphous Magnet", Phys. Letters, A68 (1978) 403-6 (Mag).

1396. S.J.POON and J.DURAND, "Magnetic-Cluster Description of Spin Glasses in Amorphous La-Gd-Au Alloys", Phys. Rev., B18 (1978) 6253-64 (Mag).

1397. S.J.POON and W.L.JOHNSON, "Electronic and Transport Properties of Liquid Quenched Amorphous $(Mo_{1-x}Ru_x)_{80}P_{20}$ Superconductors", J. Physique, 39 (1978) C6-451-3 (S).

1398. N.A.PRATTEN and M.G.SCOTT, "Stability of Some Metalloid-Free Metallic Glasses", Scripta Met., 12 (1978) 137-42 (M).

1399. B.PREDEL and G.DUDDEK, "On a Novel Technique for Preparation of Metastable Solids by Extremely Rapid Quenching of Liquid Alloys", Z. Metallkde., 69 (1978) 773-6 (in German) (QT).

1400. B.PREDEL and K.HÜLSE, "Metastable Phases of the Aluminium-Magnesium System", Z. Metallkde., 69 (1978) 690-6 (in German) (MCS, TSSE, T).

1401. S.J.RAEBURN and R.V.ALDRIDGE, "The Hall Effect, Resistivity and Magnetic Moment of Amorphous and Polycrystalline Iron Films", J. Phys. F: Metal Phys., 8 (1978) 1917-28 (VD; E, Mag).

1402. K.RAJ, J.DURAND, J.I.BUNDNICK and S.SKALSKI, "NMR Studies of Metal and Metalloid Site Hyperfine Field Distributions in Magnetically Ordered Amorphous Alloys", J. Appl. Phys., 49 (1978) 1671-6 (AQS).

1403. K.V.RAO, R.MALMHÄLL, G.BÄCKSTRÖM and S.M.BHAGAT, "Critical Behavior of an Amorphous Ferromagnet - Effects of Thermal Cycling and Annealing", Ann. Isr. Phys. Soc., 2 (1978) 435-8 (Mag)(see Chem Abstr., 89(1978) 121805p).

1404. V.V.RAO, S.P.MUKHERJEE and R.KUMAR, "Structure and Mechanical Properties of Rapidly Solidified Al Alloys", Trans. Indian Inst. Metals, 31 (1978) 383-7 (M).

1405. O.RAPP, J.E.GRINDBORG and K.V.RAO, "Saturation of Low Temperature Electrical Resistance in the Amorphous $(Fe_{1-x}Ni_x)_{75}P_{16}$-$B_6Al_3$ Alloy System", J. Appl. Phys., 49 (1978) 1733-4 (E).

1406. O.RAPP. B.LINDBERG, H.S.CHEN and K.V.RAO, "New Disordered Superconductors", J. Less-Common Metals, 62 (1978) 221-4 (S).

1407. R.RAY and R.HASEGAWA, "Rapidly Quenched Metastable Iron-Boron Solid Solutions", Solid State Commun., 27 (1978) 471-4 (TSSE, Mag).

1408. J.J.RAYMENT and B.CANTOR, "Splat Quenching of High Speed Tool Steels", Metal Sci., 12 (1978) 156-63 (AQS).

1409. M.D.RECHTIN, J.B.VANDERSANDE and P.M.BALDO, "Ion-Implantation Damage in Amorphous and Crystalline $Nb_{40}Ni_{60}$", Scripta Met., 12 (1978) 639-43.

1410. J.RICHTER, "Electronic Structure of Amorphous Alloys", Phys. Stat. Sol., (b) 87 (1978) K89-92 (AQS).

1411. E.P.ROMANOV, H.C.FREYHARDT and L.SCHULTZ, "Superconducting Composites Produced by Rapid Quenching", Scripta Met., 12 (1978) 151-6 (AQS).

1412. A.SAITO, H.KANEDA, H.NAKAMAURA, J.YAMASAKI, H.FUKUNAGA and K.NARITA, "Effect of Elastic Tensile Stress on the Static Magnetic Characteristics of Amorphous Magnetic Alloys", J. Fac. Engg. Tottori Univ., 9 (1978) 26-36 (in Japanese) (Mag)(see Chem. Abstr., 90(1979) 214248u).

1413. Y.SAKURAI, "Amorphous Magnetic Materials", J. Inst. Electron & Commun. Eng. Japan, 61 (1978) 840-9 (in Japanese) (R)(see Phys. Abstr., 80(1979) 28561).

1414. I.R.SARE and R.W.K.HONEYCOMBE, "Splat Cooling of Iron-Molybdenum-Carbon Alloys", J. Mater. Sci., 13 (1978) 1991-2002 (AQS).

1415. G.V.S.SASTRY, C.SURYANARAYANA, O.N.SRIVASTAVA and H.A.DAVIES, "Crystallization of an Amorphous Al-Pd Alloy", Trans. Indian Inst. Metals, 31 (1978) 292-4 (AB, AS).

1416. G.V.S.SASTRY, C.SURYANARAYANA, M.VAN SANDE and G.VAN TENDELOO, "A New Ordered Phase in the Al-Pd System", Mater. Res. Bull., 13 (1978) 1065-70 (MCS).

1417. E.M.SAVITSKY, K.N.IVANOVA, A.V.REVYAKIN and Yu.V.JEFIMOV, "The Formation of Non-Metallic Inclusions in High-Melting Point Metals Quenched from the Liquid State", Fiz. & Khim. Obrab Mater., No. 5 (1978) 91-7 (in Russian)(see Phys. Abstr., 82(1979) 57007); Prakt. Metallogr., 15(1978) 244-55 (AQS).

1418. E.M.SAVITSKY, Yu.V.JEFIMOV, P.MÜLLER, G.G.MUCHIN, A.V.REVYAKIN and T.M.FROLOVA, "Influence of Cooling Rate on the Structure of High-Melting Point Transition Metals", Krist. & Tech., 13 (1978) 863-72 (in German)(see Phys. Abstr., 82(1979) 43287).

1419. A.SCHÄFER and G.MENZEL, "The Structural and Electrical Properties of Metglas and Amorphous Metal Films", Thin Solid Films, 52 (1978) 11-21 (AQS, E).

1420. I.SCHMIDT and E.HORNBOGEN, "The Formation of Metastable Crystalline Phases and Glasses in Splat-Cooled Fe-C Alloys", Z. Metallkde., 69 (1978) 221-7 (AS, MCS).

1421. J.SCHNEIDER, A.HANDSTEIN and R.HESSKE, "Mictomagnetic to Ferromagnetic Transition in Amorphous Ni-Be Based Alloys", Phys. Stat. Sol., (a) 45 (1978) K47-50 (Mag).

1422. B.SCHRÖDER, W.GROßMAN, W.L.JOHNSON, C.C.TSUEI and P.CHAUDHARI, "A Comparative Study of Amorphous and Crystalline Superconducting Molybdenum Films by Ultraviolet Photoemission Spectroscopy", Solid State Commun., 28 (1978) 631-3 (VD; S).

1423. G.SCHRÖDER, R.SCHÄFER and H.KRONMÜLLER, "Magneto-Optical Investigation of the Domain Structure of Amorphous $Fe_{80}B_{20}$ Alloys", Phys. Stat. Sol., (a) 50 (1978) 475-81 (Mag).

1424. P.J.SCHURER and A.M.HORRISH, "Continuous versus Discrete Hyperfine-Field Distributions in Amorphous Ferromagnetic Iron Alloys", Solid State Commun., 28 (1978) 819-23 (Mag).

1425. M.G.SCOTT, "Crystallization Kinetics of Fe-Ni Based Metallic Glasses", J. Mater. Sci., 13 (1978) 291-6 (AB).

1426. M.G.SCOTT, M.KIJEK and H.MATYJA, "Some Comments on Microstructure of Rapidly Quenched Al-Cu Alloy", J. Mater. Sci., 13 (1978) 1354-5 (AQS).

1427. H.SEKINE, K.INOUE and K.TACHIKAWA, "Structure and Superconducting Properties of Nb-Zr Alloy Films Made by a High-Rate Sputtering", J. Japan Inst. Metals, 42 (1978) 424-32 (in Japanese) (Sp; AQS, S).

1428. V.K.SETHI, R.GIBALA and A.HEUER, "Transmission Electron Microscopy of Shear Bands in Amorphous Metallic Alloys", Scripta Met., 12 (1978) 207-9.

1429. L.Y.L.SHEN, H.S.CHEN, R.C.DYNES and J.P.GARNO, "The Energy Loss Spectra and Auger Spectra of Palladium Hydride and Palladium Glasses", J. Phys. Chem Solids, 39 (1978) 33-8 (AQS).

1430. K.SHIIKI and M.KUDO, "Production Method of Amorphous Metal Ribbons", Zairyo Kagaku, 15 (1978) 141-6 (in Japanese) (R) (see Chem. Abstr., 90(1979) 125712y).

1431. Y.SHIMADA and H.KOJIMA, "The Hall and the Magnetoresistive Effect of Amorphous Fe-Si Films", J. Appl. Phys., 49 (1978) 932-3 (Sp; Mag).

1432. Y.SHIMADA and H.KOJIMA, "Thermal Stability of Amorphous Ferromagnetic Thin Films Produced by Sputtering", Phys. Stat. Sol., (a) 47 (1978) K119-21 (Sp; T).

1433. T.SHISHIDO, K.OKAMURA and S.YAJIMA, "$Gd_3Al_5O_{12}$ Phase Obtained by Crystallization of Amorphous Gd_2O_3.5/3 Al_2O_3", J. Amer. Cer. Soc., 61 (1978) 373-5 (MCS).

1434. T.SHISHIDO, K.OKAMURA and S.YAJIMA, "Ln-M-O Glasses Obtained by Rapid Quenching Using a Laser Beam", J. Mater. Sci., 13 (1978) 1006-14 (AS).

1435. W.H.SHULL, D.G.NAUGLE, S.J.POON and W.L.JOHNSON, "Calorimetric and Resistive Measurements of Amorphous 'Splat Cooled' La_{1-x}-Ga_x Foils", Phys. Rev., B18 (1978) 3263-70 (E, T).

1436. E.SIEGEL, "Empirical Agreement of Correlation Between Superconducting Transition Temperature and Solute Concentration in Amorphous and Crystalline Binary Transition Metal Alloys: Local Superconductivity?", J. Mag. Mag. Mater., 7 (1978) 190-4 (S).

1437. V.A.SINGH and L.M.ROTH, "Spin Waves in Amorphous Ferromagnets", J. Appl. Phys., 49 (1978) 1642-4 (Mag).

1438. S.P.SINGHAL and H.HERMAN, "Neutron Small-Angle Scattering Study of Phase Decomposition in Au-Pt", J. Appl. Cryst., 11 (1978) 572-7 (AQS).

1439. S.P.SINGHAL and H.HERMAN, "Spinodal Decomposition in Amorphous Ar Implanted Pt", Appl. Phys. Letters, 32 (1978) 25-6 (AS, AB).

1440. J.SITEK, M.PREJSA, J.CIRAK, M.HUCL and J.LIPKA, "Mössbauer Study of the Amorphous $Pd_{70}Fe_{10}Si_{20}$ Alloy Upon Transition to the Crystalline State", J. Non-Cryst. Solids, 28 (1978) 225-9 (AB).

1441. F.SOMMER, G.DUDDEK and B.PREDEL, "New Glassy Alloys", Z. Metallkde., 69 (1978) 587-90 (in German) (AS).

1442. F.SPAEPEN, "A Structural Model for the Interface Between Amorphous and Crystalline Si or Ge", Acta Met., 26 (1978) 1167-77 (SMA).

1443. F.SPAEPEN, "Structural Imperfections in Amorphous Metals", J. Non-Cryst. Solids, 31 (1978) 207-21 (R).

1444. A.M.STEWART and W.A.PHILLIPS, "Electrical and Thermal Properties of Some Amorphous Metals", Phil. Mag., B37 (1978) 561-7 (E, T).

1445. M.STUBIČAR, T.IVEZIČ, E.BABIĆ and M.OČKO, "The Microhardness of Transition Metal-Metalloid Glasses", Fizika, 10 Suppl. 2 (1978) 244-9 (M)(see Phy. Abstr., 82(1979) 57085).

1446. C.SURYANARAYANA and R.C.GUPTA, "Rapid Solidification", J. Inst. Engineers (Hindi), 59, No. 8 (1978) 16-20 (in Hindi) (R).

1447. C.SURYANARAYANA, S.K.TIWARI and T.R.ANANTHARAMAN, "A New Metastable Phase in the Al-Mg System", Z. Metallkde., 69 (1978) 155-6 (MCS).

1448. H.G.SUZUKI and K.YAMAMOTO, "Mobility of Metalloid Atoms in Iron-Base Amorphous Metals Investigated with Electron Spectroscopy", Mater. Sci. & Engg., 33 (1978) 57-61 (T).

1449. K.SUZUKI, "Structure of Melt Quenched Amorphous Metals", Zairyo Kagaku, 15 (1978) 133-40 (in Japanese) (R)(see Chem. Abstr., 90(1979) 46661c).

1450. H.SZYMCZAK, "On Magnetostriction in Amorphous Thin Films", Phys. Sata. Sol., (b) 88 (1978) K97-101 (Mag).

1451. H.SZYMCZAK, "Dipolar Mechanism for Magnetostriction in Amorphous Ferromagnetic Alloys", IEEE Trans. Mag., MAG-14(1978) 847-8 (Mag).

1452. L.TAKÁCS, "Mössbauer Investigation of Ferromagnetic Metallic Glasses", Report KFKI-78-12, Hungarian Acad. Sci., Budapest (February 1978), 9pp (Mag)(see Phys. Abstr., 81(1978) 92009).

1453. M.TAKAHASHI, C.O.KIM, M.KOSHIMURA and T.SUZUKI, "Temperature Dependence of the Magnetic Moment in Rapidly Quenched Ni-B Alloys", Japan. J. Appl. Phys., 17 (1978) 741-2 (Mag).

1454. M.TAKAHASHI, C.O.KIM, M.KOSHIMURA and T.SUZUKI, "Temperature Dependence of the Saturation Magnetization in Amorphous Co-B Alloys", Japan. J. Appl. Phys., 17 (1978) 1911-2 (Mag).

1455. S.TAKAYAMA, "Recent Developments in Metallic Glasses", Bull. Japan. Inst. Metals, 17 (1978) 618-9 (in Japanese) (R).

1456. M.TAKEUCHI, "Method for Determination of Boron in Non-Crystalline Alloys", Kyotsu Shisetsu Gijutsu Kenkyu Hokoku, 8 (1978) 11-5 (in Japanese)(see Chem. Abstr., 91(1979) 32241).

1457. L.E.TANNER, "Physical Properties of Ti-Be-Si Glass Ribbons", Scripta Met, 12 (1978) 703-8 (M, E).

1458. L.E.TANNER and B.C.GIESSEN, "Structure and Formation of Metastable Phase m-TiBe", Met. Trans., A9 (1978) 67-9 (MCS).

1459. T.TARNOCZI, I.NAGY, C.HARIGITAI and M.HOSSO, "The Role of Fe_3B Compound in the Crystallization of Fe-B Glasses", IEEE Trans. Mag., MAG-14 (1978) 1025-7 (AB).

1460. T.TARNOCZI, I.NAGY and M.HOSSO, "High Temperature Magnetic Analysis of Metallic Glasses", Report KFKI-78-11, Hungarian Acad. Sci., Budapest (February 1978), 11pp. (Mag)(see Phys. Abstr., 81(1978) 91258).

1461. J.A.TARVIN, G.SHIRANE, R.J.BIRGENEAU and H.S.CHEN, "Magnetic Excitations in an Amorphous Ferromagnet", in Transition Metals 1977, eds. M.J.G.Lee, J.M.Perz and E.Fawcett (Inst. Phys., London, 1978) pp. 514-8 (Mag).

1462. J.A.TARVIN, G.SHIRANE, R.J.BIRGENEAU and H.S.CHEN, "High-Temperature Spin Dynamics in an Amorphous Ferromagnet", Phys. Rev., B17 (1978) 241-8 (Mag).

1463. N.TEOH, W.TEOH, S.ARAJS and C.A.MOYER, "Absolute Thermoelectric Power of Amorphous Metallic Glass $Fe_{80}B_{20}$ Between 300 and 1000 K", Phys. Rev., B 18 (1978) 2666-73 (T).

1464. N.TERAO, "Experimental Study of Crack Propagation in Brittle Amorphous Solids: Refraction of a Propagating Crack by Stress Waves", Phys. Stat. Sol., (a) 49 (1978) 237-48 (M).

1465. S.K.TIWARI, K.CHATTOPADHYAY, C.SURYANARAYANA and T.R.ANANTHARAMAN, "Structure of a Splat-Cooled Al-30% Mg Alloy", Bulletin EMSI, 2 (1978) 35-6 (AB, MCS).

1466. J.TOTH, "Kinetics of Crystallization of Fe-B Glass With Electrical Resistance Measurements", Report KFKI-78-33, Hungarian Acad. Sci., Budapest May 1978, 12pp; Mater. Res. Bull., 13 (1978) 691-6 (AB).

1467. J.TOTH, "Resistivity of Iron-Boron Amorphous Alloys in Magnetic Field at Low Temperatures", Report KFKI-78-13, Hungarian Acad. Sci., Budapest, February 1978, 7pp (E)(see Phys. Abstr., 81 (1978) 91667).

1468. J.TOTH, "Grange-Kiefer Transformation Applied to Metallic Glass Crystallization", Report KFKI-78-71, Hungarian Acad. Sci. Budapest, (September 1978) 15pp (in Hungarian) (AB) (see Phys. Abstr., 82(1979) 73105).

1469. C.C.TSUEI, "High Temperature Superconducting Tape Via the Amorphous State", Appl. Phys. Letters, 33 (1978) 262-3 (S).

1470. C.C.TSUEI, "Anomalous Electrical Conduction in Disordered and Non-Crystalline Metallic Conductors", Solid State Commun., 27 (1978) 691-5 (E).

1471. C.C.TSUEI, "Amorphous Superconducting Films", Proc. Electrochem. Soc., 78 (1978) 354-8 (S)(see Chem Abstr., 81(1978) 51865f).

1472. S.TSUKAHARA, T.SATOH and T.TSUSHIMA, "Magnetic Anisotropy Distribution Near the Surface of Amorphous Ribbons", IEEE Trans. Mag., MAG-14 (1978) 1022-4 (Mag).

1473. N.TSUYA and K.I.ARAI, "New Ribbon-Form Materials", Solid State Phys., 13 (1978) 237-42 (in Japanese) (R)(see Phys. Abstr., 81(1978) 92210).

1474. N.TSUYA, K.I.ARAI and T.OHSAKA, "Saturation Magnetostriction of Iron-Rich Amorphous Ribbons and Electronically Controllabale Delay Line", IEEE Trans. Mag., MAG-14 (1978) 946-8 (Mag).

1475. E.VAFAEI-MAKHSOOS, E.L.THOMAS and L.E.TOTH, "Electron Microscopy of Crystalline and Amorphous Ni-P Electrodeposited Films. In-Situ Crystallization of an Amorphous Solid", Met. Trans., A9 (1978) 1449-60 (ED; AB).

1476. B.VASVARI, "Electron Transport in Amorphous Metals", Report KFKI-78-44, Hungarian Acad. Sci. Budapest, 1978, 60 pp. (E) (see Phys. Abstr., 82 (1979) 4546).

1477. B.VASVARI, "Electrical Resistivity of Amorphous Metallic Alloys", in Transition Metals 1977. eds. M.J.G.Lee, J.M.Perz and E.Fawcett (Inst. Phys., London, 1978) pp. 210-3 (E).

1478. I.VINCZE, "Evaluation of Complex Mössbauer Spectra in Amorphous and Crystalline Ferromagnets", Solid State Commun., 25 (1978) 689-93 (AQS) (also as Report KFKI-77-71, Hungarian Acad. Sci., Budapest, September 1977, 13pp).

1479. I.VINCZE and E.BABIĆ, "Hyperfine Field Distribution in Amorphous $Fe_{40}Ni_{40}B_{20}$", Solid State Commun., 27 (1978) 1425-8 (Mag) (also as Report KFKI-78-53, Hungarian Acad. Sci., Budapest, June 1978, 10pp).

1480. I.VINCZE, T.KEMÈNY and S.ARAJS, "Quasi-Crystalline Structure of Amorphous Alloys", Report KFKI-78-54, Hungarian Acad. Sci., Budapest, June 1978, 11pp (AQS)(see Phys. Abstr., 82(1979) 4255).

1481. S.V.VONSOVSKII and E.A.TUROV, "Metallic Glasses and Amorphous Magnetism", Izv. Akad. Nauk SSSR, Ser. Fiz., 42 (1978) 1570-80 (in Russian); Bull. Acad. Sci. USSR, Phys. Ser., 42 No. 8 (1978) 1-10 (Mag)(see Phys. Abstr., 82(1979) 82818).

1482. J.L.WALTER, S.F.BARTRAM and I.MELLA, "Formation and Crystallization of Alloys with Two Amorphous Phases", Mater. Sci. & Engg., 36 (1978) 193-205 (AB).

1483. J.L.WALTER, S.F.BARTRAM and R.R.RUSSELL, "Crystallization of Amorphous Alloys $Fe_{50}Ni_{30}B_{20}$ and $Fe_{80}B_{20}$", Met. Trans., A9 (1978) 803-14 (AB).

1484. J.L.WALTER and F.E.LUBORSKY, "The Ductile-Brittle Transition of Some Amorphous Alloys", Mater. Sci. & Engg., 33 (1978) 91-4 (M).

1485. R.WANG, M.D.MERZ and J.L.BRIMHALL, "A New Amorphous $Mo_{46}Co_{54}$ Alloy Prepared by High Rate Sputter Deposition", Scripta Met., 12 (1978) 1037-41 (Sp; AS).

1486. H.WARLIMONT, "Structure and Properties of Metallic Glasses (I)", Z. Metallkde., 69 (1978) 212-20 (in German) (R).

1487. Y.WASEDA, "A Survey of the Structural Information on Amorphous Metals", J. Mater. Soc. Japan, 15 (1978) 215 (in Japanese) (R).

1488. Y.WASEDA and H.S.CHEN, "On the Structure of Amorphous $Fe_{84}B_{16}$", Solid State Commun., 27 (1978) 809-11 (AQS).

1489. Y.WASEDA and H.S.CHEN, "A Structural Study of Metallic Glasses Containing Boron", Phys. Stat. Sol., (a) 49 (1978) 387-92 (AQS).

1490. Y.WASEDA and H.S.CHEN, "Calculation of Electrical Resistivity and Its Temperature Coefficient of Amorphous $Cu_{60}Zr_{40}$ Alloy Using t-Matrix", Phys. Stat. Sol., (b) 87 (1978) 777-82 (E).

1491. Y.WASEDA and T.MASUMOTO, "Effect of Temperature and Deformation on the Structure of Amorphous Metals", Sci. Rep. RITU., 21-30 (AB).

1492. Y.WASEDA and W.A.MILLER, "Structural Relaxation in Metallic Glasses", Phys. Stat. Sol., (a) 49 (1978) K31-6 (AB).

1493. H.WATANABE, H.MORITA and H.YAMAUCHI, "Magnetic Properties of Amorphous Co-B Alloys", IEEE Trans. Mag., MAG-14 (1978) 944-5 (Mag).

1494. T.WATANABE, Y.TANABE and N.IMAMURA, "Anomalous Magnetization/ Temperature Characteristics in Electroless Ni-B Amorphous Plating Films", J. Japan Inst. Metals, 42 (1978) 562-6 (in Japanese) (Mag).

1495. L.S.WEINMAN and J.N.DEVAULT, "A Microprobe Study of Rapidly Solidified Laser Surface Alloyed Low Carbon Steels", AIP Conf. Proc., 50 (1978) 239-44 (AQS) (see Phys. Abstr., 82(1979) 70201).

1496. H.R.WENDT and F.F.ABRAHAM, "Empirical Criterion for the Glass Transition Region based on Monte-Carlo Simulations", Phys. Rev. Letters, 41 (1978) 1244-6 (T).

1497. D.B.WILLIAMS and J.W.EDINGTON, "Comments on Microstructure of Rapidly Quenched Al-Cu Alloy, Reply", J. Mater. Sci., 13 (1978) 1356-7 (AQS).

1498. C.G.WINDSOR, D.S.BOUDREAUX and M.C.NARASIMHAN, "A Search for Anisotropy in the Structure Factor of Amorphous $Pd_{0.8}Si_{0.2}$ Ribbon", Phys. Letters, A67 (1978) 282-4 (AQS).

1499. W.WOLF, "Magnetic Properties of Amorphous $(Fe, Ni)_{80}P_{14}B_6$ and $(Fe, Ni)_{83}B_{15}C_2$ Alloys", J. Mag. Mag. Mater., 9 (1978) 200-2 (Mag).

1500. E.P.WOHLFARTH, "Some Fundamental Aspects of Amorphous Magnetism", IEEE Trans. Mag., MAG-14 (1978) 933-7 (Mag).

1501. E.P.WOHLFARTH, "Magnetic Properties of Crystalline and Amorphous Alloys: A Systematic Discussion Based on the Rhodes-Wohlfarth Plot", J. Mag. Mag. Mater., 7 (1978) 113-20 (R, Mag).

1502. E.P.WOHLFARTH, "The Spatial Fluctuations of the Magnetization in Amorphous Ferromagnets", Phys. Letters, A69 (1978) 222-4 (Mag).

1503. J.V.WOOD and R.W.K.HONEYCOMBE, "Splat Quenching of Austenitic Steels Containing Niobium and Carbon", Phil. Mag., A37 (1978) 501-15 (AQS).

1504. J.V.WOOD and S.C.KING, "Production of Dental Alloy by the Melt Extraction Technique", J. Mater. Sci., 13 (1978) 1119-24 (AQS, M).

1505. M.YAGI and T.ANAYAMA, "Switching Characteristics and Domain Structures in Amorphous Toroidal Cores", IEEE Trans. Mag., MAG-14 (1978) 1019-21 (Mag).

1506 T.YAMAGUCHI and K.NARITA, "Preparation of Amorphous Magnetic Powder by Atomization Technique", Appl. Phys. Letters, 33 (1978) 468-70 (At; QT).

1507. R.YAMAMOTO and M.DOYAMA, "Computer Simulation for the Structure, Tensile Deformation and Radiation Damage of Amorphous Metals", Oyo Buturi, 47 (1978) 887-92 (in Japanese) (AQS, M)(see Phys. Abstr., 82(1979) 42137).

1508. R.YAMAMOTO, K.HAGA, H.SHIBUTA and M.DOYAMA, "Search for Bernal Holes in a Realistic Structural Model of Amorphous Iron", J. Phys. F: Metal Phys., 8 (1978) L179-82 (SMA).

1509. R.YAMAMOTO, T.HOSHIAI, H.MATSUOKA, K.TAIRA and M.DOYAMA, "The Calculations of the Heat of Crystallization and Bulk Modulus of Glassy Iron", Phys. Stat. Sol., (a) 46 (1978) 671-4 (T, M).

1510. R.YAMAMOTO, H.MATSUOKA and M.DOYAMA, "Structural Re axation of the Dense Random Packing Model for Amorphous Iron", Phys. Stat. Sol., (a) 45 (1978) 305-14 (SMA).

1511. R.YAMAMOTO, H.MATSUOKA and M.DOYAMA, "A Realistic Structural Model of Glassy Iron", Phys. Letters, A64 (1978) 457-9 (SMA).

1512. R.YAMAMOTO, T.MIHARA and M.DOYAMA, "A Chain of Bernal Holes Observed in a Realistic Model of the Amorphous Structure", J. Phys. Soc. Japan, 45 (1978) 1421-2 (SMA).

1513. R.YAMAMOTO, T.MIHARA and M.DOYAMA, "Comments on the Glass Forming Region of the Metal-Metalloid Systems Predicted by the Polk Model", Phys. Stat. Sol., (a) 50 (1978) 165-9 (SMA).

1514. H.N.YOON and A.EISENBERG, "Dynamic Mechanical Properties of Metallic Glasses", J. Non-Cryst. Solids, 29 (1978) 357-64 (M).

1515. P.G.ZIELINSKI and H.MATYJA, "Metallic Glass", Hutnik, 45 (1978) 22-36 (in Polish) (R)(see Chem. Abstr., 89(1978) 78564u).

1516. A.ZENTKO, A.ZENTKOVA and P.DUHAJ, "Magnetic Properties of Amorphous Ni-Pd-Si Alloys", J.Physique, 39 (1978) C6-951-2 (Mag).

'RAPIDLY QUENCHED METLAS III',(Proceedings of the Third International Conference on Rapidly Quenched Metals, ed. B.Cantor, The Metals Society, London 1978)

1517. T.R.ANANTHARAMAN, P.RAMACHANDRARAO, C.SURYANARAYANA, S.LELE, K.CHATTOPADHYAY, G.V.S.SASTRY and H.A.DAVIES, "Rapid Quenching of Aluminium Alloys", 1, 126-37 (R).

1518. A.S.ARGON and H.Y.KUO, "The Structure of, and Plastic Flow in a Bragg Bubble Glass", 2, 269-76 (SMA, M).

1519. E.BABIĆ, Ž.MAROHNIĆ, D.PAVUNA and B.LEONTIĆ, "Resistometric Analysis of the Phase Transformations in Amorphous $Fe_{40}Ni_{40}B_{20}$ Alloys", 1, 355-62 (E, AB).

1520. I.C.BAIANU, "Structural Studies of $Co_{0.9}P_{0.1}$ Non-Crystalline Solids Prepared by Electroless Deposition", 2 419-24 (CD; AQS).

1521. Z.BASHIR and M.R.MACKLEY, "Rapid Cooling Processing Techniques Applied to Blends of Polyethylene and Paraffin Wax", 2, 213-24 (PM).

1522. H.BECK and R.OBERLE, "On the Stability of Metallic Glasses", 1, 416-9 (SMA).

1523. B.S.BERRY and W.C.PRITCHET, "Thermally-Activated Internal Friction Peaks in Amorphous Films of Nb_3Si", 2, 21-8 (M).

1524. J.BEVK, "Superconducting Properties of Liquid-Quenched Nb_3(Al, Si) and Nb_3(Al, Ge)", 2, 17-20 (S).

1525. K.H.J.BUSCHOW and N.M.BEEKMANS, "Magnetic and Electric Properties of Amorphous Alloys of Gd and C, Al, Ga, Ni, Cu, Rh or Pd", 2, 133-6 (Mag, E).

1526. G.S.CARGILL III, "Microstructural Variability and Magnetic Anisotropy in Amorphous Gd-Co-Based Sputtered Films", 2, 364-71 (Mag).

1527. G.S.CARGILL III and C.C.TSUEI, "Atomic Scale Structure of Amorphous Sputtered Nb_3Ge and Nb_3Si", 2, 337-44 (AQS).

1528. C.CHAUMONT, H.LAVILLE, J.B.MONTE L and J.C.BERNIER, "Rapid Quenching of Ferrimagnetic Oxides", 1, 401-5 (AQS).

1529. R.CLAMPITT, M.G.SCOTT, K.L.AITKEN, L.GOWLAND and P.D.PREWETT, "Electric Field Atomisation from the Melt", 1, 57-62 (QT).

1530. E.W.COLLINGS, C.E.MOBLEY, R.E.MARINGER and H.L.GEGEL, "Selected Properties of Melt-Extracted Titanium-base Polycrystalline Alloys", 1, 188-92 (M, PP).

1531. E.W.COLLINGS, C.E.MOBLEY, R.E.MARINGER and H.L.GEGEL, "Effect of Changing Boron and Silicon Contents on Selected Properties of $Fe_{80}(P_{13}C_7)$, $Fe_{80}(P_{13}C_7)_{20-y}Mt_y$ Alloys", 2, 285-92 (PP, T, M).

1532. S.M.COPLEY, M.BASS, E.W.VAN STRYLAND, D.G.BECK and O.ESQUIVEL, "Microstructures of Surface Alloyed Ag-Cu Films Produced by Laser Melt Quenching", 1, 147-50 (AQS).

1533. P.J.COTE and L.V.MEISEL, "Amorphous Metals : Unique Test Material for Electron Transport Theory", 2, 62-8 (SMA, E).

1534. A.R.COX and E.C.VAN REUTH, "Potential Uses of Rapidly Solidified Alloys in Gas Turbine Engines", 2, 225-31 (A).

1535. S.D.DAHLGREN, "Vapour-Quenching Techniques", 2, 36-47 (VD; R).

1536. H.A.DAVIES, "Rapid Quenching and Formation of Metallic Glasses", 1, 1-21 (R).

1537. S.DAVIS, M.FISCHER, B.C.GIESSEN and D.E.POLK, "Formation and Properties of Refractory Metal Glasses II: T_5-T_9 Glasses (T_5=Nb, Ta; T_9=Rh, Ir)", 2, 425-30 (AS, T, M).

1538. J.P.DISMUKES and G.J.SELLERS, "Metallic Glasses for Improved Magnetic Shielding", 2, 205-12 (Mag).

1539. M.DOI, H.KOSAKI and T.IMURA, "Structure, Stability and Mechanical Properties of Liquid-Quenched Fe-based and Ni-based Amorphous Alloys", 2, 372-9 (AQS, M).

1540. I.W.DONALD and H.A.DAVIES, "The Formation and Stabilities of Some Iron and Nickel Based Glassy Alloys", 1, 273-80 (AQS).

1541. J.W.DRIJVER, A.L.MULDER, W.C.EMMENS and S.RADELAAR, "Stability and Mechanical Properties of Metglas 2826A", 1, 363-70 (M).

1542. F.DUFLOS and B.CANTOR, "Martensite in Splat-Quenched Iron-Nickel", 1, 110-8 (AQS).

1543. P.DUHAJ, P.MRAFKO and P.BUTVIN, "A Contribution to the Interpretation of the Structure of Pd-Si Amorphous Alloys", 1, 347-54 (AQS, AB).

1544. J.DURAND, C.THOMPSON and A.AMAMOU, "Magnetic and Transport Properties of Amorphous Ferromagnetic $(Fe_{100-x}M_x)_{79}P_{13}B_8$ Alloys Obtained by Splat Cooling (M=Mn, Cr, V; $1\leq x\leq 4$), 2, 109-19 (Mag, E).

1545. D.DUŽEVIĆ and O.ESQUIVEL, "Examination of Some Rapid Quenched Cobalt-Tungsten Carbide Alloys", 1, 151-4 (AQS).

1546. T.EGAMI, R.S.WILLIAMS and Y.WASEDA, "Structure of Amorphous Alloys Studied by Energy Dispersive X-Ray Diffraction Method", 2, 318-24 (AQS).

1547. J.E.EVETTS, W.HOWARTH and M.R.J.GIBBS, "The Micromagnetic Properties of Amorphous Alloy", 2, 127-32 (Mag).

1548. H.FRANKE, U.HEROLD, U.KÖSTER and M.ROSENBERG, "Metastable Phases in Rapidly Quenched Fe-B Alloys", 1, 155-62 (AS, MCS).

1549. H.FUJIMORI, T.KATO, T.MASUMOTO and H.MORITA, "Iron-Rich Amorphous Soft Ferromagnets with High B_s", 2, 240-4 (Mag).

1550. H.FUJIMORI, S.OHTA, T.MASUMOTO and K.NAKAMOTO, "Magnetic After-Effect on Soft Magnetic Properties of Amorphous Ferromagnets", 2, 232-9 (Mag).

1551. K.FUKAMICHI, M.KIKUCHI, H.HIROYOSHI and T.MASUMOTO, "Anomalous Thermal Expansion, ΔE Effect, Invar and Elinvar Characteristics of Some Fe-based Amorphous Alloys", 2, 117-26 (PP, M, Mag).

1552. T.FUKUNAGA, M.MISAWA, K.FUKAMICHI, T.MASUMOTO and K.SUZUKI, "Compositional Study on Short-Range Structure of Pd-Si and Fe-B Amorphous Alloys", 2, 325-32 (AQS).

1553. P.H.GASKELL, "A Model for Amorphous Palladium-Silicon and Similar Glasses with Defined Local Coordination", 2, 277-84 (SMA).

1554. B.C.GIESSEN and R.O.ELLIOTT, "Properties of Metallic Glasses Containing Actinide Metals: I.Thermal Properties of U-M Glasses (M=V, Cr, Mn, Fe, Co and Ni) 1, 406-11 (AS).

1555. B.C.GIESSEN, J.HONG, L.KABACOFF, D.E.POLK, R.RAMAN and R.St.AMAND, "Compositional Dependence of the Thermal Stability and Related Properties of Metallic Glasses I.Tg for $Ca_{65}M_{35}$ and $Zr_{475}Cu_{475}M_{05}$ Glasses" 1, 249-60.

1556. C.D.GRAHAM, Jr and T.EGAMI, "Magnetic Properties of Amorphous Materials", 2, 109-16 (R).

1557. N.J.GRANT, "Applications of Metastable Microcrystalline Alloys", 2, 172-82 (R).

1558. W.A.GRANT, A.ALI, L.T.CHADDERTON, P.J.GRUNDY and E.JOHNSON, "Production of Amorphous Alloys by Ion Implantation", 1, 63-72 (II; AS).

1559. A.L.GREER and J.A.LEAKE, "The Stability and Transformations of $Fe_{80}B_{20}$ Metallic Glasses", 1, 299-306 (AB).

1560. R.HAILES, J.H.VINCENT and H.JONES, "Structure of Al-3 to 18Wt-%Ni Gun-Quenched from the Melt", 1, 163-6 (AQS).

1561. S.HATTA, T.EGAMI and C.D.GRAHAM,Jr, "Development of Amorphous Alloys With High Saturation Induction", 2, 183-7 (Mag).

1562. M.V.HEIMENDAHL and G.MAUSSNER, "Kinetics of Crystallization Studied by T.E.M. in Metglas 2826A ($Fe_{32}Ni_{36}Cr_{14}P_{12}B_6$)", 1, 428-30 (AB).

1563. U.HEROLD and U.KÖSTER, "The Influence of Metal or Metalloid Exchange on Crystallization of Amorphous Iron-Boron Alloys", 1, 281-90 (AB).

1564. H.HILMANN and H.R.HILZINGER, "On the Formation of Amorphous Ribbons by The Melt Spin Technique", 1, 22-9 (PM).

1565. H.HILMANN and H.R.HILZINGER, "Correlation Between the Temperature Coefficient of the Electrical Resistivity and the Crystallization Temperature of Amorphous Fe-Ni-B Alloys", 1, 371-4 (T, E)

1566. R.W.K.HONEYCOMBE, "Rapidly Quenched Crystalline Alloys", 1, 73-84 (R).

1567. E.HORNBOGEN and I.SCHMIDT, "Metastable Crystalline Phases and Glasses in Ternary Fe-C-X Alloys", 1, 261-4 (AS, AB, MCS).

1568. K.HOSELITZ, "Magnetic Iron-Silicon-Boron Metallic Glass", 2, 245-8 (Mag, M).

1569. O.HUNDERI and K.LONVIK, "Thermosonimetric Investigation of the Crystallization of Metallic Glasses", 1, 375-82 (AB).

1570. A.INOUE, T.MASUMOTO, S.ARAKAWA and T.IWADACHI, "Mechanical Properties and Thermal Stability of New Type Amorphous High-Carbon Alloy Steels", 1, 265-72 (M, T).

1571. T.IVEZIČ, M.OČKO, E.BABIĆ, M.STUBIČAR and Ž.MAROHNIĆ, "Correlation Between Microhardness and Magnetic Properties in Amorphous Fe-Ni Alloys", 2, 380-3 (M, Mag).

1572. W.L.JOHNSON, "Metastable Superconducting Alloys", 2, 1-16 (R).

1573. N.S.KAZAMA, M.MITERA and T.MASUMOTO, "The Size Effect of Metalloids on the Magnetic Properties of Iron based Amorphous Alloys", 2, 164-71 (Mag).

1574. T.KEMÈNY, I.VINCZE, B.FOGARASSY and S.ARAJS, "On the Stability and Crystallization of Fe-B Alloys", 1, 291-8 (AB).

1575. M.KIRITANI, T.YOSHIIE and F.E.FUJITA, "Diffusion and Crystallization in Vacuum-Deposited,Liquid-Quenched Amorphous Alloys", 2, 308-17 (AB).

1576. M.LASOCKA and H.MATYJA, "Glass Transition and Stability Range in $Te_{80}(GeA^{IV})_{20}$ Glasses Obtained by Rapid Cooling of Melt", 1, 239-48 (T).

1577. B.LEONTIĆ, J.LUKATELA, E.BABIĆ and M.OČKO, "Low Temperature Quenching Techniques", 1, 41-8 (QT).

1578. B.G.LEWIS, H.A.DAVIES and K.D.WARD, "The Crystallization of $(Fe_{100-x}Ni_x)_{83}B_{17}$ Amorphous Alloys", 1, 325-36 (AB).

1579. M.LIARD, M.MÜLLER, K.P.ACKERMANN, B.DELLEY, K.AGYEMAN, H.U.KÜNZI, H.RUDIN and H.-J.GÜNTHERODT, "Preparation and Physical Propeties of Metallic Glasses", 2, 29-35 (M, Mag, E).

1580. H.H.LIEBERMANN, "Gas Boundry Layer Effects in Proceeding Glassy Alloy Ribbons", 1, 34-40 (PM).

1581. J.LIVAGE, N.GHARBI, M.MICHAUD and A.REVCOLEVSCHI, "Fibrous Texture of Amorphous Semi-Conducting V_2O_5", 1, 412-5 (AQS).

1582. T.C.LONG, T.A.DONELLY, J.DURAND and D.G.ONN, "Low Temperature Specific Heat of the Metallic Glasses $(Ni_{99}Fe_1)_{79}P_{13}B_8$, $(Fe_{98}Cr_2)_{79}P_{13}B_8$ and $(Fe_{96}Cr_4)_{79}P_{13}B_8$", 2, 69-72 (T).

1583. F.E.LUBORSKY, J.J.BECKER and H.H.LIEBERMANN, "Replacement of Boron by in Fe-B-C Amorphous Alloys", 2, 249-52 (Mag).

1584. F.E.LUBORSKY, H.H.LIEBERMANN, J.J.BECKER and J.L.WALTER, "The Magnetic Properties of Fe-B Amorphous Alloys", 2, 188-96 (Mag).

1585. R.MALMHÄLL, G.BÄCKSTRÖM, K.V.RAO and S.M.BHAGAT, "Hall Resistivity of Fe and Fe-Ni based Alloys", 2, 145-8 (Mag).

1586. G.MARCHAL, Ph.MANGIN, M.PIECUCH,B.RODMACQ and Chr.JANOT, "Electrical Resistivity of Amorphous Fe_xSn_{1-x} Alloys", 2, 73-80 (VD; E).

1587. R.E.MARINGER and C.E.MOBLEY, "Advances in Melt Extraction", 1, 49-56 (QT).

1588. Ž.MAROHNIĆ, E.BABIĆ, J.IVKOV and A.HAMZIĆ, "Galvanomagnetic Effects in $Fe_xNi_{80-x}B_{20}$ and $Fe_xNi_{80-x}P_{14}B_6$ Metallic Glasses", 2, 149-55 (E).

1589. T.B.MASSALSKI, U.MIZUTANI, K.T.HARTWIG and R.W.HOPPER, "Low Temperature Specitic Heat Measurements on Amorphous Alloy Ribbons", 2, 81-8 (T).

1590. T.MASUMOTO, K.HASHIMOTO and M.NAKA, "Corrosion Behaviour of Amorphous Metals", 2, 435-48 (R).

1591. J.MEGUSAR, A.S.ARGON and N.J.GRANT, "Deformation Localization, Plastic Instabilities and Fracture in a $Pd_{80}Si_{20}$ Glass", 2, 392-400 (M).

1592. T.MIZOGUCHI, T.KUDO and T.IRISAWA, "The Structure of a Metal-Metal Amorphous Alloy $Cu_{0.57}Zr_{0.43}$ Determined by Neutron Diffraction", 2, 384-91 (AQS).

1593. P.MÜLLER, E.M.SAVITSKY, Yu.V.JEFIMOV and V.MATTHES, "Superconductivity of Ternary Alloys with A15-Structure in as Splat Cooled Conditions", 2, 89-95 (S).

1594. T.MURATA, T.MASUMOTO and M.SAKAI, "Slip Deformation and Critical Shear Stress of Amorphous Pd-Si Alloy", 2, 401-4 (M).

1595. M.NAGUMO, T.TAKAHASHI, T.ARAI and T.HASEGAWA, "Thermal Stability of Ductility in Amorphous Metallic Alloys", 1, 383-6(M).

1596. M.NAKA, K.HASHIMOTO, K.ASAMI and T.MASUMOTO, "Corrosion Behaviour of Amorphous Cobalt-Base Alloys", 2, 449-56 (C).

1597. M.NAKA, Y.NISHI and T.MASUMOTO, "Critical Cooling rate for Glass Formation of Pd-Cu-Si Alloys", 1, 231-8 (PM).

1598. S.NANAO, Y.OHJI, J.SUGIURA and H.INO, "Mössbauer Effect and Electrical Resistivity in Splat-Quenched La-Au Alloys", 2, 48-55 (E, S).

1599. H.NEUHAUSER, "High-Speed Cinematography of Shear Band Formation in Metallic Glasses", 2, 431-4 (M).

1600. J.NIEWIAROWSKI and H.MATYJA, "Splat Quenching of High-Speed Tool Steels", 1, 193-7 (AQS).

1601. W.B.NOWAK and M.F.COLLINS, "Corrosion Resistance of Ion-Plated Metallic Films", 2, 475-64 (C).

1602. S.OHNUMA and T.MASUMOTO, "Amorphous Magnetic Alloys (Fe, Co, Ni)-(Si, B) with High Permeability and Its Thermal Stability", 2, 197-204 (Mag).

1603. S.N.OJHA, K.CHATTOPADHYAY and P.RAMACHANDRARAO, "Solidification of Eutectic Alloys over a Heat Pipe", 1, 119-25 (AQS).

1604. J.P.PATTERSON, A.L.GREER, J.A.LEAKE and D.R.H.JONES, "Hot Forming of a Metallic Glass", 2, 293-300 (F).

1605. D.E.POLK, C.E.DUBE and B.C.GIESSEN, "The Effect of Oxygen Additions on the Properties of Amorphous Transition Metal Alloys", 1, 220-30 (T).

1606. R.B.POND, Sr. and C.R.SIRIAN, "Techniques for Determinging Very Short Room Temperature Lifetimes of Amorphism Developed by Rapid Quenching", 2, 301-7 (PM).

1607. N.A.PRATTEN and M.G.SCOTT, "Annealing Effects in Metal-Metal Amorphous Alloys", 1, 387-91 (M).

1608. P.RAMACHANDRARAO, "Glass Transition Interval in Metallic Glasses", 1, 392-6 (T).

1609. J.J.RAYMENT and B.CANTOR, "Splat-Quenched Tungsten Steels", 1, 85-93 (AQS).

1610. J.RIVORY and B.BOUCHET, "Influence of Heat Treatments on the Optical Properties of Amorphous CoP and NiP", 2, 56-61 (PP).

1611. P.N.ROSS and B.H.KEAR, "Surface Segregation in Rapidly Solidified Superalloys", 1, 102-9 (AQS).

1612. J.F.SADOC and A.LIÉNARD, "Cu-Zr and Large Atoms-Small Atoms Alloys Structure", 2, 405-9 (SMA).

1613. E.SAHIN and H.JONES, "Extended Solid Solubility, Grain Refinement and Age-Hardening in Al- 1 to 13wt-%Zr Rapidly Quenched From the Melt", 1, 138-46 (AQS, AB).

1614. E.M.SAVITSKY, Yu.V.JEFIMOV, G.G.MUCHIN and T.M.FROLOVA, "Structure and Properties of Equilibrium and Rapidly Quenched Cu-Containing Alloys", 1, 167-70 (AQS).

1615. A.S.SCHAAFSMA, H.SNIJDERS and F.VAN DER WOUDE, "Crystallization of Amorphous $Fe_{80}B_{20}$", 1, 428-30 (AB).

1616. M.G.SCOTT, "Thermal Stability and Crystallization of Metallic Glasses", 1, 198-213 (R).

1617. P.H.SHINGU, K.SHIMOMURA, R.OZAKI, K.OSAMURA and Y.MURAKAMI, "Crystallization of Amorphous Fe-P-C and Fe-Ni-P-B Alloys", 1, 315-24 (AB).

1618. F.SPAEPEN, "Structure and Flow of Amorphous Alloys", 2, 253-68 (R).

1619. P.R.STRUTT, H.NOWOTNY, YOUNG-Won KIM and B.H.KEAR, "Structural Analyses of Laser Surface Melted Tool Steels", 1, 171-5 (AQS).

1620. K.SUZUKI, F.ITOH, T.FUKUNAGA and T.HONDA, "A Comparision of Atomic and Electronic Structures Between Electodeposited and Melt-Quenched Ni-P Amorphous Alloys", 2, 410-4 (AQS).

1621. A.M.TONEJC and A.KIRIN, "X-ray Diffraction Study of Splat-Quenched Ag-Sn Alloys", 1, 176-9 (AQS).

1622. E.TÖRÖK and G.HAUSCH, "Magnetoelastic Effects of Some Ferromagnetic Metallic Glasses", 2, 105-8 (Mag).

1623. H.G.WAGNER, U.GONSER and A.SCHERTZ, "Investigation of the Structure of Amorphous $Fe_{80}B_{20}$", 2, 333-6 (AQS).

1624. J.L.WALTER, "A Study of the Formation of Amorphous Ribbon Using High-Speed Motion Pictures", 1, 30-3 (PM).

1625. J.L.WALTER and S.F.BARTRAM, "Crystallization of Some Amorphous Alloys", 1, 307-14 (AB).

1626. R.WANG, M.D.MERZ, J.L.BRIMHALL and S.D.DAHLGREN, "A Refractory Glassy Alloy of $W_{50}Fe_{50}$", 1, 420-3 (AS).

1627. Y.WASEDA, "A Survey of the Current Structural Information on Metallic Glasses", 2, 352-63 (SMA).

1628. Y.WASEDA and H.S.CHEN, "The Structure of Amorphous Metal-Metal Alloys", 2, 415-8 (AQS).

1629, R.S.WILLIAMS and T.EGAMI, "Temper Embrittlement of Amorphous Alloys", 1, 214-9 (M).

1630. J.WONG, F.W.LYTLE, R.B.GREGOR, H.H.LIEBERMANN, "Local Structure and Dynamic Disorder of Fe & Ni in Metallic Glasses by EXAFS Spectroscopy", 2, 345-51 (AQS).

1631. J.V.WOOD, J.K.BINGHAM and J.V.BEE, "Rapid Solidification of High Alloy Steels and Nickel Superalloys", 1, 94-101 (AQS).

1632. J.V.WOOD, A.CROSSLEY and W.M.STOBBS, "Rapid Solidification and Ageing of Cu-Al-Ni Marmem Alloys", 1, 180-7 (AQS).

1633. Z.WROŃSKI, J.SUWALSKI and H.MATYJA, "Mössbauer Effect Study of Structural Relaxation and of Early Stages of Crystallization in $Fe_{80}B_{20}$ Glass", 1, 397-400 (AQS, AB).

1634. K.ZAVETA, J.SCHNEIDER, A.HANDSTEIN and F.ZOUNOVA, "Superparamagnetic Behaviour of $Fe_{1.6}Ni_{78.4}B_{10}P_{10}$ Metallic Glass", 2, 156-63 (Mag).

1635. P.G.ZIELIŃSKI, J.OSTATEK, M.KIJEK and H.MATYJA, "Formation and Stability of Metallic Glasses Containing Zirconium", 1, 337-46 (AS, T).

METALLIC GLASSES (American Society for Metals, Metals Park, Ohio, U.S.A., 1978)(All are review articles)

1636. R.ALBEN, J.I.BUDNICK and G.S.CARGILL III, "Magnetic Structures", 304-39 (Mag).

1637. B.S.BERRY, "Elastic and Anelastic Behavior", 161-89 (M).

1638. H.S.CHEN and K.A.JACKSON, "The Influence of Alloy Composition on Glass Formation and Properties", 74-96 (PM).

1639. L.A.DAVIS, "Strength, Ductility and Toughness", 190-223 (M).

1640. J.DIXMIER and J.F.SADOC, "Structural Models", 97-113 (SMA).

1641. H.-J.GÜNTHERODT and H.U.KÜNZI, "Electronic Properties of Metallic Glasses and Melts", 247-74 (E).

1642. E.M.GYORGY, "Magnetic Properties", 275-303 (Mag).

1643. S.KAVESH, "Principles of Fabrication", 36-73 (QT).

1644. J.C.M.LI, "Micromechanisms of Deformation and Fracture", 224-46 (M).

1645. D.E.POLK and B.C.GIESSEN, "Overview of Principles and Applications", 1-35 (PM).

1646. F.SPAEPEN and D.TURNBULL, "Atomic Transport and Tranformation Behavior", 114-27 (T).

1647. D.R.UHLMANN and R.W.HOPPER, "Relaxation in Glasses", 128-60 (AQS).

STRUCTURE AND PROPERTIES OF AMORPHOUS METALS, eds. T.Masumoto and K.Suzuki (Suppl. Sci. Rep. RITU, A27 (1978)).

1648. K.I.ARAI and N.TSUYA, "Magnetostriction of Fe-Rich Amorphous Ribbons", pp. 215-9 (in Japanese) (Mag).

1649. H.FUJIMORI and T.MASUMOTO, "Soft Magnetic Properties of Amorphous Fe, Co, Ni-Base Alloys with Si and B", pp. 181-97 (Mag).

1650. F.E.FUJITA, "Mössbauer Spectroscopy of Amorphous Alloys", pp. 95-103 (AQS).

1651. K.FUKAMICHI, M.KIKUCHI, H.HIROYOSHI and T.MASUMOTO, "Invar and Elinvar Characteristics in Amorphous Fe-Base Alloys", 199-214 (M, PP).

1652. K.HASHIMOTO, "Corrosion Behavior of Amorphous Alloys", pp. 233-50 (C).

1653. T.IMURA, M.DOI and H.KOSAKI, "Determination and Identification of Atomic Structure of Amorphous Alloys", pp. 47-62 (AQS).

1654. H.INO and S.NANAO, "A Study of Local Atomic and Electronic Structure of Amorphous Alloys by Mössbauer Spectroscopy", pp. 105-16 (AQS).

1655. T.KATAYAMA and T.TSUSHIMA, "Physical Origins of the Perpendicular Anisotropy in Amorphous Rare Earth-Transition Metal Films", pp. 159-69 (Sp; Mag).

1656. N.S.KAZAMA, N.HEIMAN and H.WATANABE, "Recent Development in Amorphous Magnetism", pp. 131-57 (Mag, R).

1657. M.KIRITANI and T.YOSHIIE, "Diffusion and Crystallization in Vacuum-Deposited and Liquid-Quenched Amorphous Alloys", pp. 63-72 (AB, T).

1658. K.MAEDA and S.TAKEUCHI, "Elementary Process of Deformation of Amorphous Metals", pp. 87-94 (M).

1659. T.MIZOGUCHI, "Basic Magnetic Properties of Amorphous Alloys", pp. 117-29 (Mag).

1660. K.MURAKAMI, "Applications of Amorphous Magnetic Cores to Instrument and Control Devices", pp. 221-31 (A).

1661. P.H.SHINGU and R.OZAKI, "The Formation of Amorphous Metals by Liquid Quenching", pp. 21-35 (AS, PM).

1662. K.SUZUKI, "Structure and Vitrification in Amorphous Metals", pp. 1-20 (R, AQS).

1663. S.TSUKAHARA, T.SATOH and T.TSUSHIMA, "Magnetic Anisotropy Distributed Near the Surface of Amorphous Ribbons", pp. 171-80 (Mag).

1664. N.TSUYA and K.I.ARAI, "Amorphous Ribbons of $LiNbO_3$ and $PbTiO_3$", pp. 43-6 (AS).

1665. R.YAMAMOTO and M.DOYAMA, "Computer Simulations of the Structure and the Tensile Deformation of Amorphous Iron", pp. 73-86 (AQS, M).

RAPID SOLIDIFICATION PROCESSING : PRINCIPLES AND TECHNOLOGIES, eds. R.Mehrabian, B.H.Kear and M.Cohen (Claitor's Publishing Division, Baton Rouge, La, U.S.A., 1978)

1666. G.J.ABBASCHIAN, "Kinetic Considerations during Rapid Melting and Solidification of Surface Layers", pp. 84-6 (PM).

1667. E.H.AIGELTINGER, S.R.BATES, R.W.GOULD and J.J.HREN, "Phase Equilibria in Rapidly Solidified Ni-Rich Ni-Mo-Al Alloys", pp. 291-305 (AQS).

1668. D.BECK, S.M.COPLEY, M.BASS and E.VAN STRYLAND, "Solidification of Superalloy Melts Produced by Laser Irradiation", pp. 104-11 (AQS).

1669. R.G.BOURDEAU and J.B.MOORE, "Thermomechanical Processing of RSR Superalloy Powders", pp. 334-46 (F).

1670. E.M.BREINAN and B.H.KEAR, "Rapid Solidifaction Laser Processing of Materials for Control of Microstructure and Properties", pp. 87-103 (QT).

1671. W.L.BROWN, J.A.GOLOVCHENKO, K.A.JACKSON, L.C.KIMERLING, H.J.LEAMY, G.L.MILLER, J.M.POATE, J.W.RODGERS, G.A.ROZGONYI, T.T.SHENG, T.N.C.VENKATESAN and G.K.CELLER, "Laser Annealing of Ion-Implanted Semiconductors", pp. 123 8 (II, AB).

1672. R.W.CAHN, "Rapid Solidification by Plasma Spraying", pp.129-39 (R).

1673. W.S.CEBULAK, "Processing/Property Interactions in High Strength P/M Mill Products", pp. 324-33 (F).

1674. C.F.CLINE, J.MAHLER, M.FINGER, W.KUHL and R.W.HOPPER, "Explosive Fabrication of Rapidly Solidified Alloys", pp.380-7 (F).

1675. H.A.DAVIES, B.G.LEWIS and I.W.DONALD, "Comparison of Roller Quenching and Melt Spinning for Continuous Casting of Strip", pp. 78-83 (PM).

1676. E.J.DULIS and J.H.MOLL, "Properties of HIP Consolidated Superalloy and Titanium Alloys", pp. 362-79.

1677. R.D.FIELD and H.L.FRASER, "Microstructural Observations of Metal Powders Using Analytical Electron Microscopy", pp. 270-7 (AQS).

1678. M.FISCHER, D.E.POLK and B.C.GIESSEN, "Refractory Metallic Glasses Prepared by an Arc Furnace Quenching Unit", pp. 140-8 (QT, AS).

1679. A.M.GLASS and K.NASSAU, "Twin-Roller Quenched Ferroelectric Oxide Glasses", pp. 149-53 (AS).

1680. M.R.GLICKSTEIN, R.J.PATTERSON II and N.E.SHOCKEY, "Rapid Solidification Effects on Micron-Sized Droplets", pp. 46-63 (AQS).

1681. N.J.GRANT, "A Review of Various Atomization Processes", pp. 230-45 (QT).

1682. P.R.HOLIDAY, A.R.COX and R.J.PATTERSON II, "Rapid Solidification Effects on Alloy Structures", pp. 246-57 (AQS).

1683. H.JONES, "Some Principles of Solidification at High Cooling Rates", pp. 28-45 (PM).

1684. S.KAVESH, "Metallic Glasses - The Technology of Rapid Solidification Processing", pp. 165-87 (R).

1685. A.LAWLEY, "Processing-Structure-Property Relationships in Consolidated Powders", pp. 306-23.

1686. D.J.LOOFT and E.C.VAN REUTH, "Rapid Solidification Processing: An Overview", pp. 1-8 (R).

1687. R.E.MARINGER and C.E.MOBLEY, "The Melt Extraction of Metallic Filament, Staple Fiber and Powder", pp. 208-21 (QT).

1688. R.MEHRABIAN, "Relationship of Heat Flow to Structure in Rapid Solidification Processing", pp. 9-27 (PM).

1689. T.E.MILES and J.F.RHODES, "Unique Powder Processing for Ultimate Shape and Properties", pp. 347-61 (F).

1690. C.E.MOBLEY, R.E.MARINGER and L.DILLINGER, "Surface Features of Rapidly Quenched Filaments", pp. 222-9 (AQS).

1691. J.PEREL, J.F.MAHONEY, B.E.KALENSHER, K.E.VICKERS and R.MEHRABIAN, "Electrohydrodynamic Generation of Submicron Particles for Rapid Solidification", pp. 258-69 (QT).

1692. H.REISS and J.L.KATZ, "Nucleation in Rapid Solidification Processes", pp. 64-77 (PM).

1693. S.R.ROBERTSON, T.J.GORSUCH and R.P.I.ADLER, "Advanced Melt Extraction Technology", pp. 188-207 (QT).

1694. P.N.ROSS and B.H.KEAR, "Surface Segregation in Rapidly Solidified Superalloy Powders", pp. 278-84 (AQS).

1695. R.J.SALZBRENNER, "Some Preliminary Studies on RSP 9Ni-4Co Steels", pp-285-90 (AQS).

1696. A.R.E.SINGER, "Spray Deposition for the Manufacture of Thin Strip", pp. 154-64 (QT).

1697. M.TULI, P.R.STRUTT, H.NOWOTNY and B.H.KEAR, "Laser Surface Melting of High Speed Tool Steels", pp. 112-6 (AQS).

1698. L.S.WEINMAN and C.KIM, "Rapid Solidification of Aluminium and Titanium Alloy Surfaces", pp. 117-22 (AQS).

1979

1699. K.AGYEMEN, R.MULLER and C.C.TSUEI, "Alloying Effect on Superconductivity in Amorphous La-Based Alloys", Phys. Rev., B19 (1979) 193-8 (S).

1700. D.AKHTAR, V.D.VANKAR, T.C.GOEL and K.L.CHOPRA, "Splat Quenching of Lead in Vacuum", J. Mater. Sci., 14 (1979) 983-7 (AQS, MCS).

1701. D.AKHTAR, V.D.VANKAR, T.C.GOEL and K.L.CHOPRA, "Metastable Structures of Liquid-Quenched and Vapour-Quenched Antimony", J. Mater. Sci., 14 (1979) 988-94 (AQS).

1702. D.AKHTAR, V.D.VANKAR, T.C.GOEL and K.L.CHOPRA, "Stabilization and Transformation Kinetics of the Metastable Phases of Liquid-Quenched Antimony", J. Mater. Sci., 14 (1979) 2422-6 (AB).

1703. D.AKHTAR, V.D.VANKAR, T.C.GOEL and K.L.CHOPRA, "Metastable Structures of Splat-Cooled and Vapour-Deposited Pb and Sb Films", Thin Solid Films, 58 (1979) 327 (AQS).

1704. D.ALIAGA GUERRA, J.DURAND, W.L.JOHNSON and P.PANISSOD, "NMR Study on ^{31}P in an Amorphous Superconducting $(Mo_{0.5}Ru_{0.5})_{80}$-P_{20} Alloy", Solid State Commun., 31 (1979) 487-91 (S).

1705. P.ALLIA and F.VINAI, "Structural Instabilities and Magnetic Properties of Amorphous Ferromagnets", J. Appl. Phys., 50 (1979) 7662-4 (Mag).

1706. A.AMAMOU, "Mössbauer Effect and Short Range Order in Fe-P-B and Fe-P-C Amorphous Alloys", Phys. Stat. Sol., (a) 54 (1979) 565-72 (AQS).

1707. A.AMAMOU and G.KRILL, "Photoemission and Electronic Structure of Crystalline Co_3B and Amorphous Co-P-B", Solid State Commun., 31 (1979) 971-5 (AQS).

1708. T.R.ANANTHARAMAN, "The Magic of Metallic Glasses", Trans. Indian Inst. Metals, 32, No. 1 (1979) xii-xix (R)(see Chem. Abstr., 92 (1980) 80533p).

1709. P.M.ANDERSON III, J.STEINBERG and A.E.LORD, Jr., "Continuous Cooling Curves (CT) Versus Isothermal Transformation (TTT) Diagrams in Metallic Alloy Glasses", J. Non-Cryst. Solids, 34 (1979) 267-72 (PM).

1710. T.R.ANTHONY and H.E.CLINE, "A Fluid-Flow-Temperature Model for the Casting of Amorphous Metal Ribbon by Melt Extraction", J. Appl. Phys., 50 (1979) 239-44 (PM).

1711. T.R.ANTHONY and H.E.CLINE, "Dimensional Variations in Newtonian-Quenched Metal Ribbons Formed by Melt Spinning and Melt Extraction", J. Appl. Phys., 50 (1979) 245-54 (PM).

1712. K.I.ARAI, N.TSUYA, K.OHMORI and Y.AOSHIMA, "Magnetic Properties of Ribbon-Form Sendust", Trans. Inst. Electron & Commun. Eng. Japan., E62, No. 6 (1979) 440 (Mag)(see Phys. Abstr., 83(1980) 15761).

1713. H.ARAKI, G.PARK, A.HIKATA and C.ELBAUM, "Ultrasonic Attenuation and Velocity Studies of Amorphous PdSiCu", Solid State Commun., 32 (1979) 625-9 (PP).

1714. A.S.ARGON, "Plastic Deformation in Metallic Glasses", Acta Met., 27 (1979) 47-58 (M).

1715. A.S.ARGON, G.W.HAWKINS and H.Y.KUO, "Reinforcement of Mortar with Metglas Fibers", J. Mater. Sci., 14 (1979) 1707-16 (M).

1716. A.S.ARGON and H.Y.KUO, "Plastic Flow in a Disordered Bubble Raft (an Analog of a Metallic Glass)", Mater. Sci. & Engg., 39 (1979) 101-9 (M).

1717. G.R.ARMSTRONG and H.JONES, "Effect of Decreased Section Thickness on the Formation, Structure and Properties of a Chill-Cast Aluminium-Silicon Alloy", in Solidification and Casting of Metals (The Metals Society, London, 1979) pp.454-9 (AQS, PM).

1718. D.G.AST and D.J.KRENITSKY, "Stress Relaxation in Coiled Ribbons of $Fe_{40}Ni_{40}P_{14}B_6$ and $Fe_{29}Ni_{49}P_{14}B_6Si_2$", J. Mater. Sci., 14 (1979) 287-95 (M).

1719. A.AUDORD, J.C.JOUSSET and J.DURAL, "30K Fast Neutron and Fission Fragment Irradiations of Two Fe-B Type Amorphous Alloys", Radiation Effects Letters, 50 (1979) 9-14 (see Phys. Abstr., 83(1980) 19846).

1720. T.AYANO, H.OHNO, K.KAWAMURA and K.FURUKAWA, "Structural Analysis of Amorphous $Pd_{0.8}Si_{0.2}$ Alloy by Dense Random Racking Model of Differently Sized Compressible Spheres", Phys. Stat. Sol., (a) 51 (1979) 325-32 (SMA).

1721. T.AYANO, H.OHNO, K.UTUNI, K.KAWAMURA and K.FURUKAWA, "Irradiation Effects of Helium Ions on the Structure of Amorphous $Pd_{0.8}Si_{0.2}$ Alloy", J. Nucl. Mater., 79 (1979) 430-1.

1722. N.AZAM, L.LE NAOUR, C.RIVERA, P.GROSJEAN, P.SACOVY and J.DELAPLACE, "Crystallization and Effects of Irradiation in an Amorphous Alloy of Type $Fe_{40}Ni_{38}Mo_4B_{18}$", J. Nucl. Mater., 83 (1979) 298-304 (AB) (in French) (AB)(see Phys. Abstr., 82 (1979) 90558).

1723. R.A.BAARTMAN, B.HEINRICH, J.F.COCHRAN and G.C.CHI, "Transmission of 24 GHz Radiation Through Co-P from 4 K to 414 K", J. Appl. Phys., 50 (1979) 7674-6 (PP).

1724. E.BABIĆ, Ž.MAROHNIĆ, F.HAJDU, M.TEGZE and I.VINCZE, "Resistivity Minimum in Amorphous and Crystalline $Fe_{40}Ni_{40}B_{20}$ Alloys", Solid State Commun., 29 (1979) 175-9 (E) (also as Report KFKI-78-74, Hungarian Acad. Sci., Budapest, October 1978, 12pp).

1725. I.C.BAIANU, J.PATTERSON and K.A.RUBINSON, "Ferromagnetic Resonance Observations of Surface Effects, Magnetic Ordering and Inhomogeneous Anisotropy in Metallic Glass", Mater. Sci. & Engg., 40 (1979) 273-84 (Mag).

1726. I.C.BAIANU, K.A.RUBINSON and J.PATTERSON, "The Observation of Structural Relaxation in FeNiPB Glass by Ferromagnetic Resonance and X-Ray Scattering", Phys. Stat. Sol., (a) 53 (1979) K133-8 (AB).

1727. M.N.BAIBICH, W.B.MUIR, G.BÉLANGER, J.DESTRY, H.S.ELZINGA and P.A.SCHROEDER, "Thermopower of Amorphous Metals from 4.2 K to 300 K", Phys. Letters, A73 (1979) 328-30 (T).

1728. S.BANERJEE and B.CANTOR, "Martensitic Transformations in Splat-Quenched Zirconium Alloys", in Proc. Internat. Conf. Martensitic Transformations, ICOMAT 1979, ed. M.Cohen, pp.195-201(AQS).

1729. J.M.BARANDIARAN, "Influence of Torsion on the Magnetic Properties of an Amorphous Ribbon", J. Phys.:D:Appl. Phys., 12 (1979) 1943-50 (Mag).

1730. S.BASAK, R.CLARKE and S.R.NAGEL, "Pair Distribution Function and Its Relation to the Glass Transition in an Amorphous Alloy", Phys. Rev., B20 (1979) 3388-90 (T, AQS).

1731. S.BASAK, R.CLARKE and S.R.NAGEL, "Temperature Dependence of the Structure Factor in Nb-Ni Glasses", Phys. Rev., B20 (1979) 4278-86 (AQS).

1732. F.BAUMANN, "Superconductivity of Amorphous Metals", Rev. Chemie Minerale, 16 (1979) 328-48 (R, S)(see Chem. Abstr., 92(1980) 86355d).

1733. A.E.BERKOWITZ, J.D.LIVINGSTON, B.D.NATHAN and J.L.WALTER, "Thermomagnetic Behavior of a Duplex Ferromagnetic Alloy, $Fe_{75}Si_{15}B_{10}$", J. Appl. Phys., 50 (1979) 1754-6 (Mag).

1734. A.BERRADA, J.DURAND, N.HASSANAIN and B.LOEGEL, "Anisotropy versus Exchange in Amorphous Heavy Rare-Earth Based RE_4Au Alloys", J. Appl. Phys., 50 (1979) 7621-3 (Mag).

1735. B.S.BERRY and W.C.PRITCHET, "Influence of the Magnetomechanical Pole Effect on the Behaviour of Thin Vibrating Reed", J. Appl. Phys., 50 (1979) 1630-2 (Mag).

1736. S.M.BHAGAT, M.L.SPANO and K.V.RAO, "Temperature Dependence of the Magnetization of Amorphous Alloys", J. Appl. Phys., 50 (1979) 1580-2 (Mag).

1737. J.L.BLACK and P.FULDE, "Influence of the Superconducting State upon the Low-Temperature Properties of Metallic Galsses", Phys. Rev. Letters, 43 (1979) 453-6 (S).

1738. E.BLASBERG, D.KORN and H.PFEIFLE, "Electrical Resistivity of Amorphous Alloy Films: SnAu", J. Phys. F: Metal Phys., 9 (1979) 1821-32 (VD; E).

1739. M.BLUNDELL, K.J.OVERSHOTT and C.D.GRAHAM, Jr., "Power Loss Measurements in Amorphous Alloys", J. Appl. Phys., 50 (1979) 1598-1600 (Mag).

1740. N.BONANOS and E.LILLEY, "Preparation of Suzuki Phases and Their Stability", Mater. Res. Bull., 14 (1979) 1609-15 (MCS).

1741. J.A.BORDERS, "Metastable Phases in Metals Produced by Ion Implantation", Annual Rev. Mater. Sci., 9 (1979) 313-39 (R) (see Met. Abstr., 13(1980) 12-0241).

1742. P.G.BOSWELL, "Solidification Models for High Cooling Rates", Metals Forum, 2 (1979) 40-54 (PM).

1743. P.G.BOSWELL, "The Wear Resistance of a Liquid-Quenched Metallic Glass", J. Mater. Sci., 14 (1979) 1505-7 (M).

1744. P.G.BOSWELL and G.A.CHADWICK, "The Structure of Splat-Quenched Cd-Zn Alloys", J. Mater. Sci., 14 (1979) 1269-79 (AQS).

1745. B.BOUCHER and B.BARBARA, "Magnetic After-Effect in the Bulk Amorphous Alloy $Tb_{52}Ag_{48}$", J. Phys. F : Metal Phys., 9 (1979) 151-9 (Mag).

1746. J.L.BRIMHALL, L.A.CHARLOT and R.WANG, "Irradiation Effects in Amorphous and Crystalline, Sputter Deposited Mo-Ni", Scripta Met., 13 (1979) 217-9 (Sp; AB).

1747. M.BROUHA and J.VAN DER BORST, "The Effect of Annealing Conditions on the Magneto-Mechanical Properties of Fe-B-Si Amorphous Ribbons", J. Appl. Phys., 50 (1979) 7594-6 (Mag, M).

1748. A.BRUNSCH, "Non-Magnetostrictive Amorphous Soft Magnetic Thin Film Material of High Magnetization", J. Appl. Phys., 50 (1979) 7600-2 (Mag).

1749. A.BRUNSCH, "Magnetic Properties and Corrosion Resistance of $(CoFeB)_{100-x}Cr_x$ Thin Films", J. Appl. Phys., 50 (1979) 7603-5 (Sp; Mag, C).

1750. K.H.J.BUSCHOW, "Crystallization and Magnetic Properties of Amorphous Gd-Fe and Er-Fe Alloys", J.Less-Common Metals, 66 (1979) 89-97 (AB, Mag).

1751. K.H.J.BUSCHOW and N.M.BEEKMANS, "On the Thermal Stability and the Electrical Resistivity of Amorphous Th-Fe Alloys", Phys. Stat. Sol., (a) 56 (1979) 505-11 (T, E).

1752. K.H.J.BUSCHOW and N.M.BEEKMANS, "Thermal Stability and Electronic Properties of Amorphous Zr-Co and Zr-Ni Alloys", Phys. Rev., B19 (1979) 3843-9 (T, E).

1753. K.H.J.BUSCHOW and N.M.BEEKMANS, "Formation, Decomposition and Electrical Transport Properties of Amorphous Hf-Ni and Hf-Co Alloys", J. Appl. Phys., 50 (1979) 6348-52 (AS, AB, E).

1754. K.H.J.BUSCHOW and W.W.VAN DEN HOOGENHOF, "Note on the Formation and Magnetic Properties of Amorphous Eu-Ag Alloys", J. Mag. Mag. Mater., 12 (1979) 123-6 (AS, Mag).

1755. K.H.J.BUSCHOW and A.M.VAN DER KRAAN, "Magnetic Properties of Amorphous $Th_{1-x}Fe_x$ Alloys", Phys. Stat. Sol., (a) 53 (1979) 665-9 (Mag).

1756. B.T.A.CHANG and J.C.M.LI, "Indentation Recovery of Amorphous Materilas", Scripta Met., 13 (1979) 51-4 (M).

1757. K.CHATTOPADHYAY and P.RAMACHANDRARAO, "Metastable Phase Formation and Decomposition in a Rapidly Solidified Al-Pt Alloy", Mater. Sci. & Engg., 38 (1979) 7-17 (AB).

1758. P.CHAUDHARI, A.LEVI and P.STEINHARDT, "Edge and Screw Dislocations in an Amorphous Solid", Phys. Rev. Letters, 43 (1979) 1517-20 (SMA).

1759. H.S.CHEN and Y.WASEDA, "Structure of Glassy Zr-Cu and Nb-Ni Alloys", Phys. Stat. Sol., (a) 51 (1979) 593-9 (AQS).

1760. Y.N.CHEN and T.EGAMI, "Effect of Annealing on the Curie Temperature of Amorphous Alloys", J. Appl. Phys., 50 (1979) 7615-7 (Mag).

1761. G.C.CHI and G.S.CARGILL III, "Structural Characterization of Amorphous Electrodeposited Co-P Alloy", J. Appl. Phys., 50 (1979) 2713-20 (ED; AQS).

1762. M.C.CHI and T.EGAMI, "The Spin Correlation in a Ferromagnet with Random Anisotropy", J. Appl. Phys., 50 (1979) 1651-3 (Mag).

1763. C.L.CHIEN, "Mössbauer Study of an Amorphous Magnetic Solid Containing Cr", Phys. Rev., B19 (1979) 81-6 (Mag).

1764. C.L.CHIEN and H.S.CHEN, "Effects of Mo on the Magnetic Properties of Amorphous $(Fe_xMo_{1-x})_{75}P_{16}B_6Al_3$", J. Appl. Phys., 50 (1979) 1574-6 (Mag).

1765. C.L.CHIEN and H.S.CHEN, "Correlation of Isomer Shift and Hyperfine Field in Amorphous $(Fe_{0.2}Co_{0.8})_{80}P_{17}Al_3$", J. Physique, 40 (1979) C2-118-9 (Mag).

1766. C.L.CHIEN, J.H.HSU, J.P.STOKES, A.N.BLOCH and H.S.CHEN, "Magnetic Ordering Temperature, Hyperfine Field Distribution and Electrical Resistivity of Amorphous $(Fe_xMn_{1-x})_{75}P_{16}B_6Al_3$", J. Appl. Phys., 50 (1979) 7647-9 (Mag, E).

1767. C.L.CHIEN, D.MUSSER, E.M.GYORGY, R.C.SHERWOOD, H.S.CHEN, F.E.LUBORSKY and J.L.WALTER, "Magnetic Properties of Amorphous Fe_xB_{100-x} $(72 \leq x \leq 86)$ and Crystalline Fe_3B", Phys. Rev., B20 (1979) 283-95 (Mag).

1768. C.L.CHIEN, D.MUSSER, F.E.LUBORSKY and J.L.WALTER, "Mössbauer Studies of Amorphous $(Fe_xNi_{1-x})_{80}B_{20}$", J. Physique, 40 (1979) C2-129-31 (Mag).

1769. C.P.CHOU, L.A.DAVIS and R.HASEGAWA, "Elastic Constants of Fe(Ni, Co)-B Glasses", J. Appl. Phys., 50 (1979) 3334-7 (M).

1770. M.H.COHEN and G.S.GREST, "Liquid-Glass Transition, A Free-Volume Approach", Phys. Rev., B20 (1979) 1077-98 (T).

1771. E.COLEMAN, "Equilibrium Crystalline Phases in Annealed (Co, Fe)$_{0.75}$B$_{0.25}$ Metallic Glasses", Mater. Sci. & Engg., 39 (1979) 261-6 (AB).

1772. P.C.COLTER, T.W.ADAIR III and D.G.NAUGLE, "Hall Effect in Amorphous $La_{1-x}Ga_x$ Foils", Phys. Rev., B20 (1979) 2959-60 (Mag).

1773. M.A.CONTINENTINO and N.RIVIER, "Magnetic Excitations in Amorphous Ferromagnetis", Phys. Stat. Sol., (b) 93 (1979) 721-33 (Mag).

1774. V.E.COSSLETT and D.J.SMITH, "High Resolution Imaging of Amorphous Materials and Crystalline Defects", Chem. Scripta, 14 (1979) 39-46 (AQS).

1775. P.J.COTE and L.V.MEISEL, "Thermopower and Resistivity of Amorphous and Crystalline NiP", Phys. Rev., B20 (1979) 3030-5 (T, E).

1776. N.COWLAM, "Magnetic Structures in Amorphous Solids", Physics Bull., 30, No. 11 (1979) 479 (R).

1777. N.COWLAM, M.SAKATA and H.A.DAVIES, "A Neutron Diffraction Study of $Fe_{83}B_{17}$ Metallic Glass", J. Phys. F : Metal Phys., 9 (1979) L203-8 (AQS).

1778. F.CUMBRERA, J.J.MALANDAIN and P.VIGIER, "Inelastic and Elastic Electronic Diffusion by Amorphous Sample", Compt. Rend., B288 (1979) 383-5 (in French) (T).

1779. H.A.DAVIES, "The Formation, Stability and Properties of Metallic Glasses", Revue de Chimie Minerale, 16 (1979) 349-69 (R).

1780. L.A.DAVIS, "Fracture Toughness of Metallic Glasses", Met. Trans., A10 (1979) 235-40 (M).

1781. F.J.A.DEN BROEDER and J.VAN DER BORST, "Magnetization Reversal in $Fe_{80}B_{15}Si_5$ Metallic Glass with Large Uniaxial Magnetostrictive Anisotropy", J. Appl. Phys., 50 (1979) 7116-21 (Mag).

1782. F.J.A.DEN BROEDER and J.VAN DER BORST, "$Fe_{80}B_{20-x}Si_x$ Glasses: A Study of Some Physical Properties as Depending on Metalloid Content", J. Appl. Phys., 50 (1979) 4279-82 (T, M, Mag).

1783. R.B.DIEGLE, "Localised Corrosion of Amorphous Fe-Ni-Cr-PB Alloys", Corrosion, 35 (1979) 250-8 (C).

1784. S.B.DIERKER, H.GUDMUNDSSON and A.C.ANDERSON, "Resistance Minima in Amorphous PdSi Alloy", Solid State Commun., 29 (1979) 767-8 (E).

1785. G.DIETZ, H.BESTGEN and G.GODLEWSKY, "Magnetic Anisotropy of Electrodeposited Co-P and Fe-P Alloys", J. Mag. Mag. Mater., 13 (1979) 71-3 (ED; Mag).

1786. A.G.DIRKS, "Amorphous to Crystalline Transition in Transition Metal-Boron Thin Films", in Phase Transformations (Institution of Metallurgists, London, 1979) Part III, pp. 30-2 (AB)(see Chem. Abstr., 92(1980) 26814t).

1787. A.G.DIRKS and J.R.M.GIJSBERS, "Crystallization of Amorphous Rare-Earth-Iron and Transition Metal-Boron Thin Films", Thin Solid Films, 58 (1979) 333-7 (AB).

1788. V.I.DOBATKIN, V.I.ELAGIN, G.A.MUDRENKO and R.M.SIZOVA, "Structure of Monotectic Alloy Systems Accompanying Rapid Cooling of Metals", Izv. Akad. Nauk. SSSR, Met., No.1 (1979) 105-11 (in Russian); Russian Met., No. 1 (1979)(see Phys. Abstr., 82(1979) 70053).

1789. K.DOI, T.AYANO and K.KAWAMURA, "Structral Changes of Amorphous $Pd_{80}Si_{20}$ with Neutron Irradiation, with an Interpretation in terms of the Imperfections in the Structure of Amorphous Solids", J. Non-Cryst. Solids, 34 (1979) 405-18 (AQS, AB).

1790. M.DOMANN, H.GRIMM and H.KRONMÜLLER, "The High-Field Magnetization Cruve of Amorphous Ferromagnetic Alloys", J. Mag. Mag. Mater., 13 (1979) 81-4 (Mag).

1791. E.R.DOMB, C.A.MACDONALD and W.L.JOHNSON, "Magnetic Impurities in Amorphous Superconductors : $T_{Kondo} > T_c$", Solid State Commun., 30 (1979) 775-8 (S).

1792. T.A.DONNELLY, T.EGAMI and D.G.ONN, "Low Temperature Specific Heat of the Metallic Glasses $Fe_xNi_{80-x}P_{14}B_6$ with x=0,20, 40, 60, 80", Phys. Rev., B20 (1979) 1211-20 (T).

1793. A.J.DREHMAN and W.L.JOHNSON, "Effects of Thermal Relaxation on Amorphous Superconducting Zr-Rh Alloy", Phys. Stat. Sol. (a) 52 (1979) 499-507 (S).

1794. G.DUBLON, C.H.LIN and J.BEVK, "Magnetization and Magnetoresistance Study of Amorphous $Fe_xPd_{82-x}Si_{18}$", J. Appl. Phys., 50 (1979) 7650-2 (Mag).

1795. J.DURAND, D.ALIAGA GUERRA, P.PANISSOD and R.HASEGAWA, "Hyperfine Field Distribution at ^{59}Co in Amorphous $(Fe, CO)_{79}P_{13}B_8$ Alloys", J. Appl. Phys., 50 (1979) 7668-70 (Mag).

1796. J.DURAND and S.J.POON, "Spin Glasses in Amorphous and Crystalline RE Alloys", J. Physique, 40 (1979) C2-231-6 (Mag).

1797. P.DUWEZ, "Metallic Glasses - A New Class of Materilas : Their Scientific and Industiral Importance", Proc. Indian Acad. Sci., C2 (1979) 117-32 (R).

1798. T.EGAMI, "Structure of Magnetic Amorphous Alloys Studied by Energy Dispersive X-Ray Diffraction", J. Appl. Phys., 50 (1979) 1564-9 (AQS, R).

1799. M.FÄHNLE, "Localized Spin Excitation in Amorphous Ferromagnets", J. Mag. Mag. Mater., 13 (1979) 50-2 (Mag).

1800. M.FÄHNLE, "Distribution of Dipole Fields in Disordered Crystalline and Amorphous Ferromagnetic Alloys", Appl. Phys., 20 (1979) 275-82 (Mag).

1801. P.H.FANG, "Analysis of Crystallization Kinetics of Amorphous PdSi Alloys", Japan. J. Appl. Phys., 18 (1979) 189-90 (AB).

1802. R.FERRER, M.J.ZUCKERMANN, A.K.BHATTACHARJER and B.COQBLIN, "Amorphous Rare Earth Alloys : Random Anisotropy Ferromagnetism in the Pair Approximation", Phys. Letters, A70 (1979) 231-4 (Mag).

1803. A.FERT and R.ASOMOZA, "Transport Properties of Magnetic Amorphous Alloys", J. Appl. Phys., 50 (1979) 1886-91 (Sp;T,R).

1804. J.L.FINNEY, "Comment on 'Short Range Structure of Amorphous Alloys'", Nature, 280 (1979) 847-8 (SMA)(see also entry 2103).

1805. V.FRANETOVIĆ, O.MILAT, D.IVČEK and A.BONEFAČIĆ, "Quenching Efficiency of Some Splat-Cooling Devices", J. Mater. Sci., 14 (1979) 498-500 (PM).

1806. Th.FRECHEN, G.DIETZ and F.GAUS, "Time Effects in Ferromagnetic Amorphous Fe-P Alloys", J. Mag. Mag. Mater., 13 (1979) 85-8 (Mag).

1807. R.L.FREED and J.B.VANDER SANDE, "The Effects of Devitrification on the Mechanical Properties of $Cu_{46}Zr_{54}$ Metallic Glass", Met. Trans., A10 (1979) 1621-6 (AB, M).

1808. H.FUJIMORI and N.S.KAZAMA, "Preparation of Bulk Amorphous Magnetic Alloys by High Rate Sputter Deposition", Solid State Phys., 14 (1979) 399-405 (in Japanese) (Sp; Mag, R)(see Chem. Abstr., 92(1980) 32397a).

1809. F.E.FUJITA, "Mössbauer Effect Study of Structure of Amorphous Alloys", J. Physique, 40 (1979) C2-120-2 (AQS).

1810. T.FUJIWARA, "Electronic States and Transport in Amorphous and Liquid Transition Metals; Fe, Co and Ni", J. Phys. F : Metal Phys., 9 (1979) 2011-24 (T).

1811. T.FUJIWARA and Y.TANABE, "Electronic States in Amorphous and Liquid Iron", J. Phys. F : Metal Phys., 9 (1979) L73-6 (AQS).

1812. T.FUJIWARA and Y.TANABE, "Electronic States in Amorphous Metals", J. Phys. F : Metal Phys., 9 (1979) 1085-95 (SMA).

1813. K.FUKAMICHI, "Amorphous Invar and Elinvar Alloys", Kinzoku, 49, No. 9 (1979) 21-8 (in Japanese) (Mag, M, R).

1814. K.FUKAMICHI, H.HIROYOSHI, M.KIKUCHI and T.MASUMOTO, "Invar Effects in Some Iron-Base Amorphous Alloys", J. Mag. Mag. Mater., 10 (1979) 294-9 (Mag).

1815. K.FUKAMICHI, M.KIKUCHI, T.MASUMOTO and M.MATSUURA, "Magnetic Properties and Invar Effects of Amorphous Gd-Co Ribbons", Phys. Letters, A73 (1979) 436-8 (M, Mag).

1816. K.FUKAMICHI, T.MASUMOTO and M.KIKUCHI, "Invar and Elinvar Amorphous Alloys", IEEE Trans. Mag., MAG-15 (1979) 1404-9 (R).

1817. M.FUKAMICHI, K.HOSHIMOTO and H.YOSHIDA, "Structure of Amorphous $Pd_{60}Ni_{20}Si_{20}$ Alloys", Scripta Met., 13 (1979) 807-11 (AQS).

1818. N.FUNAKOSHI, "Comment on Analysis of Crystallization Kinetics of Amorphous PdSi Alloys", Japan. J. Appl. Phys., 18 (1979) 1843-4 (AB).

1819. N.FUNAKOSHI, T.KANAMORI, T.MANABE and H.TOMISHIMA, "The Crystallization Kinetics of Amorphous PdAgSi Alloys", Japan. J. Appl. Phys., 18 (1979) 1885-9 (AB).

1820. P.H.GASKELL, "On the Structure of Simple Inorganic Amorphous Solids", J. Phys. C : Solid State Phys., 12 (1979) 4337-68 (R).

1821. P.H.GASKELL, "A New Structural Model for Amorphous Transition Metal Silicides, Borides, Phosphides and Carbides", J. Non-Cryst. Solids, 32 (1979) 207-24 (SMA).

1822. P.H.GASKELL, D.J.SMITH, C.J.D.CATTO and J.R.A.CLEAVER, "Direct Observation of the Structure of a Metallic Alloy Glass", Nature, 281 (1979) 465-7 (AQS).

1823. J.A.GERBER, S.G.CORNELISON, W.L.BURMESTER and D.J.SELLMYER, "Magnetic Properties of the Rare-Earth Glasses $(R_{65}Fe_{35})_{100-x}B_x$", J. Appl. Phys., 50 (1979) 1608-10 (Mag).

1824. M.R.J.GIBBS, J.E.EVETTS and N.J.SHAH, "The Effect of Mechanical Deformation on the Magnetic Properties of Amorphous Alloys", J. Appl. Phys., 50 (1979) 7642-4 (Mag).

1825. J.M.GIBSON and A.HOWIE, "Investigation of Local Structure and Composition in Amorphous Solids by High Resolution Electron Microscopy", Chem. Scripta, 14 (1979) 109-16 (AQS).

1826. B.C.GIESSEN, J.C.BARRICK and L.E.TANNER, "Formation and Structure of a New η Phase $Ti_5Be_3(O, N)_x$", Mater. Sci. & Engg., 38 (1979) 211-6 (MCS).

1827. J.J.GILMAN, "Ferrous Metallic Glasses", Metal Progress, 116, No. 7 (1979) 42-7 (R).

1828. U.GONSER, M.GHAFARI, H.G.WAGNER and H.FISCHER, "Spin Texture Determination in Amorphous $Fe_{80}B_{20}$ by Linearly Polarized Recoil-Free Gamma Rays", J. Physique, 40 (1979) C2-126-8 (Mag).

1829. J.M.GORDON, "A Polymer Model for Metal Alloy Glasses", J. Phys. Chem., 83 (1979) 889-92 (SMA)(see Phys. Abstr., 83 (1980) 6361).

1830. M.GOTO, H.TANGE and T.TOKUNAGA, "Densities of Amorphous Alloy Systems $(Fe_{1-x}M_x)_{77}Si_{10}B_{13}$ (M=Co, Ni)", Japan. J. Appl. Phys., 18 (1979) 2023-4 (PP).

1831. T.GOTO, "Microstructure and Tensile Strength of Silver-Copper Alloy Filaments Produced by Glass-Coated Melt Spinning", Trans. Japan Inst. Metals, 20 (1979) 377-82 (AQS, M).

1832. C.D.GRAHAM, Jr. and T.EGAMI, "Magnetic Properties of Amorphous Ribbons", IEEE Trans. Mag., MAG-15 (1979) 1398-1403 (R, Mag).

1833. A.L.GREER and J.A.LEAKE, "Structural Relaxation and Cross-Over Effect in a Metallic Glass", J. Non-Cryst. Solids, 33 (1979) 291-7 (AB).

1834. G.S.GREST and S.R.NAGEL, "Effective Field Distributions and Resistivity Minima in Amorphous Ferromagnets", Phys. Rev., B19 (1979) 3571-9 (E, Mag).

1835. M.GRIMSDITCH, A.MALOZEMOFF and A.BRUNSCH, "Standing Spin Waves Observed by Brillouin Scattering in Amorphous Metallic Fe_{80}-B_{20} Films", Phys. Rev. Letters, 43 (1979) 711-4 (PP).

1836. P.J.GRUNDY, A.ALI, C.E.CHRISTODOULIDES and W.A.GRANT, "Amorphous Transition Metal-Metalloid Alloy Films Prepared by Ion Implantation", Thin Solid Films, 58 (1979) 253-8 (II; AS, Mag).

1837. G.R.GRUZALSKI, J.A.GERBER and D.J.SELLMYER, "Electron Properties and Superconductivity of Zr-Pd Glasses", Phys. Rev., B19 (1979) 3469-75 (S).

1838. G.R.GRUZALSKI and D.J.SELLMYER, "Magnetically Dilute Metallic Glasses. I. 3d Moments", Phys. Rev., B20 (1979) 184-93 (Mag).

1839. G.R.GRUZALSKI and D.J.SELLMYER, "Magnetically Dilute Metallic Glasses. II. 4f Moments", Phys. Rev., B20 (1979) 194-203 (Mag).

1840. H.-J.GÜNTHERODT, "Metallic Glasses", Metall, 33 (1979) 723-6 (in German) (R).

1841. H.-J.GÜNTHERODT, H.BECK, P.OELHAFEN, K.P.ACKERMANN, M.LIARD, M.MÜLLER, H.U.KÜNZI, H.RUDIN and K.AGYEMAN, "Metallic Glasses ; Bulk and Surface Properties", in Electrons in Disordered Metals and at Metallic Surfaces, eds. P.Phariseau, B.L.Gyorffy and L.Scheire (Plenum Press, New York, 1979), pp. 501-27 (R).

1842. J.HAFNER and L.v.HEIMENDAHL, "Microscopic Calculations of the Stability of Metallic Glasses", Phys. Rev. Letters, 42 (1979) 386-9 (AS).

1843. J.HAFNER and L.v.HEIMENDAHL, "Microscopic Calculations of the Stability of Metallic Glasses", Phys. Rev. Letters, 42 (1979) 386-9 (AS, SMA).

1844. R.HASEGAWA and R.C.O'HANDLEY, "Soft Magnetic Properties of Metallic Glasses - Recent Developments", J. Appl. Phys., 50 (1979) 1551-6 (Mag, R).

1845. R.HASEGAWA and R.RAY, "Low Temperature Magnetization Study Crystalline and Glassy Fe-B Alloys", Phys. Rev., B20 (1979) 211-4 (Mag).

1846. R.HASEGAWA and R.RAY, "Magnetization of Glassy Co-B Alloys", J. Appl. Phys., 50 (1979) 1586-8 (Mag).

1847. K.HASHIMOTO, "Amorphous Stainless Alloys", Bull. Japan Inst. Metals, 18 (1979) 362-7 (in Japanese) (R)(see Chem Abstr., 92 (1980) 62914j).

1848. K.HASHIMOTO, "Corrosion Resistance of Amorphous Alloys", Corrosion Engg., 28 (1979) 352-61 (in Japanese) (R).

1849. K.HASHIMOTO, K.ASAMI, M.NAKA and T.MASUMOTO, "Surface Films Formed on Amorphous Co-Cr Alloys in 1N HCl", Corrosion Engg. (Boshoku Gijutsu), 28 (1979) 271-7 (C).

1850. K.HASHIMOTO, K.ASAMI, M.NAKA and T.MASUMOTO, "The Role of Alloying Elements in Improving the Corrosion Resistance of Amorphous Iron-base Alloys", Corr. Sci., 19 (1979) 857-67 (C).

1851. K.HASHIMOTO, M.NAKA, K.ASAMI and T.MASUMOTO, "An X-Ray Photo-Electron Spectroscopy Study of the Passivity of Amorphous Fe-Mo Alloys", Corrosion Sci., 19 (1979) 165-70 (C).

1852. S.HATTA, "High Flux Amorphous Materials for Transformers", Kinzoku, 49, No. 9 (1979) 15-20 (in Japanese) (Mag, R).

1853. S.HATTA and T.EGAMI, "Temperature Dependence of Magnetization of Amorphous Fe-B-C Alloys", J. Appl. Phys., 50 (1979) 1589-91 (Mag).

1854. S.HATTA, T.EGAMI and C.D.GRAHAM, Jr., "Fe-B-C Amorphous Alloys with Room Temperature Saturation Induction Over 17.5 kG", Appl. Phys. Letters, 34 (1979) 113-5 (Mag).

1855. E.HAUSER, R.J.ZIRKE, J.TAUC, J.J.HAUSER and S.R.NAGEL, "Optical Properties of Amorphous Metallic Au-Si Alloys", Phys. Rev., B19 (1979) 6331-6 (PP).

1856. N.HEIMAN and N.S.KAZAMA, "Magnetic Properties of Amorphous Alloys of Fe and La, Lu, Y and Zr", Phys. Rev., B19 (1979) 1623-32 (VD; Mag).

1857. N.HEIMAN and N.S.KAZAMA, "Amorphous Transition Metal Oxide Films", J. Appl. Phys., 50 (1979) 7633-5 (Sp; Mag).

1858. L.v.HEIMENDAHL, "Structure and Dynamics of a Two-Component Metallic Glass", J. Phys. F : Metal Phys., 9 (1979) 161-9 (SMA).

1859. M.v.HEIMENDAHL and G.MAUSSNER, "The Metastable Crystallization Phases in the Amorphous Alloy Metgals 2826A", J. Mater. Sci., 14 (1979) 1238-44 (AB).

1860. D.W.HENDERSON, "Thermal Analysis of Non-Isothermal Crystallization Kinetics in Glass Forming Liquids", J. Non-Cryst. Solids, 30 (1979) 301-15 (AB).

1861. H.R.HILZINGER, H.HILLMANN and A.MAGER, "Magnetostriction Measurements on Co-Base Amorphous Alloys", Phys. Stat. Sol., (a) 55 (1979) 763-9 (Mag).

1862. T.HIRATA, "On the Glass Transition Temperature in Some Metallic Glasses", Scripta Met., 13 (1979) 555-9 (T).

1863. H.H.HOHMANN, "Metallic Glasses - A New Group of Materilas", Technica, 28 (1979) 577-9 (in German) (R)(see Chem Abstr., 91 (1979) 215076u).

1864. E.HORNBOGEN, "Combined Reactions", Met. Trans., A10 (1979) 947-72 (R).

1865. K.HOSELITZ, "Crystallization of Magnetic Fe-Si-B Metallic Glasses", Phys. Stat. Sol., (a) 53 (1979) K23-6 (Mag).

1866. R.H.HOWELL and R.W.HOPPER, "Positron Life Time Study of Fe_{40}-$Ni_{40}P_{14}B_6$ Glass", Scripta Met., 13 (1979) 367-70 (AQS).

1867. G.HUNGER, K.D.WARD and H.A.DAVIES, "Annealing Embrittlement and Crystallization of Cobalt-Iron-Boron Metallic Glasses", Z. Metallkde., 70 (1979) 714-7 (AB, M).

1868. Gh. ILONCA, "Electrical Resistivity and Magnetorisistivity of Amorphous $Mn_xPd_{82-x}Ge_{18}$ Alloys", Acta Phys. Pol., A55 (1979) 275-80 (Sp; E, Mag).

1869. O.IMAMURA, "Magnetic Properties and Applications of Iron-Base Amorphous Alloy Films", Kinzoku, 49, No. 10 (1979) 27-33(Mag,R).

1870. S.INAWASHIRO, S.KATSURA and M.TAKAHASHI, "Magnetization of Amorphous Magnets", J. Physique, 40 (1979) C2-147-8 (Mag).

1871. R.INGALLS, C.M.LIU and K.V.RAO, "High Pressure Studies of Invar-Related Systems", J. Mag. Mag. Mater., 10 (1979) 257-8.

1872. H.INO, S.NANAO and T.MUTO, "A Study of Amorphous Pd-Si Alloys by Mössbauer Effect of Fe and X-Ray Diffraction", J. Phys. Soc. Japan, 46 (1979) 63-71 (AQS).

1873. Y.INOKUTI and B.CANTOR, "Martensite Morphology in Rapidly Solidified Fe Alloys", in Proc. Internat. Conf. Martensitic Transformations, ICOMAT 1979, ed. M.Cohen, pp. 46-52 (AQS).

1874. Y.INOKUTI, F.DUFLOS and B.CANTOR, "Martensite Morphology in Solidified Pure Iron", in Phase Transformations (Institution of Metallurgists, London, 1979), Part IV, pp. 17-9 (AQS)(see Phys. Abstr., 83(1980) 7547).

1875. K.INOMATA, M.HASEGAWA and S.SHIMANUKI, "Substituted Amorphous Co-Fe-Si-B Alloys", Japan. J. Appl. Phys., 18 (1979) 937-41 (Mag).

1876. A.INOUE, S.ARAKAWA and T.MASUMOTO, "Effect of Alloying Elements on Defect Structure and Hardness of $M_{23}C_6$ Type Carbides", Trans. Japan Inst. Metals, 20 (1979) 585-92 (M,AB).

1877. A.INOUE, T.IWADACHI and T.MASUMOTO, "Amorphous High-Carbon Tungsten Steels Rapidly Quenched from Melts", Trans. Japan Inst. Metals, 20 (1979) 76-8 (AQS).

1878. A.INOUE, A.KITAMURA and T.MASUMOTO, "Ni-B and Co-B Amorphous Alloys with High Boron Concentration", Trans. Japan Inst. Metals, 20 (1979) 404-6 (AS).

1879. A.INOUE, Y.KOJIMA, T.MINEMURA and T.MASUMOTO, "Formation of Ductile Ni_3Al Type Compound in Fe-(Ni, Mn)-Al-C Systems by Splat Quenching", Trans. Japan Inst. Metals, 20 (1979) 468-71 (AQS, M).

1880. A.INOUE and T.MASUMOTO, "Superconducting Properties of Amorphous Metals", Kinzoku, 49, No. 12 (1979) 24-32 (in Japanese)(R, S).

1881. A.INOUE and T.MASUMOTO, "Formation of Non-Equilibrium Cr_3C Carbide in Cr-C Binary Alloys Quenched Rapidly From the Melt", Scripta Met., 13 (1979) 711-5 (MCS, AB).

1882. A.INOUE, T.NAOHARA and T.MASUMOTO, "Crystallization Temperature and Hardness of New Type Cobalt-Based Amorphous Alloys", Trans. Japan Inst. Metals, 20 (1979) 329-36 (T, M, AS).

1883. A.INOUE, T.NAOHARA, T.MASUMOTO and K.KUMADA, "Thermal Stability and Hardness of New Type Nickel-Based Amorphous Alloys", Trans. Japan Inst. Metals, 20 (1979) 577-84 (AS, T, M).

1884. A.INOUE, S.SAKAI, H.M.KIMURA and T.MASUMOTO, "Crystallization Temperature and Hardness of New Chromium-Based Amorphous Alloys", Trans. Japan Inst. Metals, 20 (1979) 255-62 (QT, T, AS, M).

1885. J.JÄCKLE and K.FROBÖSE, "Phonon Characterization and the Theory of the Temperature Dependence of the Electrical Resistivity of Amorphous Metals", J. Phys. F : Metal Phys., 9 (1979) 967-86 (E).

1886. T.JAGIELINSKI, "Magnetostriction in Amorphous Ni-Co Alloys", J. Appl. Phys., 50 (1979) 7588-90 (Mag).

1887. E.JOHNSON, T.WOHLENBERG, W.A.GRANT, P.HANSEN and L.T.CHADDERTON, "TEM Study of Amorphous Alloys Produced by Ion Implantation", J. Microscopy, 116 (1979) 77-87 (II; AS).

1888. W.L.JOHNSON, "The Influence of Magnetic Ordering on the Superconducting Properties of Amorphous $(Zr_{1-x}Er_x)_3Rh$ Alloys", Solid State Commun., 32 (1979) 981-7 (S).

1889. W.L.JOHNSON, "Superconductivity and Electronic Properties of Amorphous Transition Metal Alloys", J. Appl. Phys., 50 (1979) 1557-63 (S, R).

1890. W.L.JOHNSON, C.C.TSUEI, S.I.RAIDER and R.B.LAIBOWITZ, "A Study of Amorphous Superconducting Transition Metal-Metalloid Alloys Produced by Co-Evaporation", J. Appl. Phys., 50 (1979) 4240-5 (VD; S).

1891. W.L.JOHNSON and A.R.WILLIAMS, "Structure and Properties of Transition Metal-Metalloid Glasses Based on Refractory Metals", Phys. Rev., B20 (1979) 1640-55 (AS, AQS).

1892. W.JUEPTNER, P.RYDER and G.SEPOLD, "Metallic Glasses as Future Construction Materials", DVS Ber., 58 (1979) 40-4 (in German) (A)(see Chem. Abstr., 92(1980) 9780m).

1893. M.KACZOROWSKI and H.MATYJA, "Decomposition of Supersaturated Solid Solution in Splat-Cooled Sn-10at.%Sb Alloy", J. Mater. Sci., 14 (1979) 2887-93 (TSSE, AB).

1894. T.KANEYOSHI, "Comment on the Indirect-Exchange Interaction in an Amorphous Magnet", Phys.Letters, A71 (1979) 287-8 (Mag).

1895. B.H.KEAR, E.M.BREINAN and L.E.GREENWALD, "Laser Glazing : A New Process for Production and Control of Rapidly Chilled Metallurgical Microstructures", in Solidification and Casting of Metals (The Metals Society, London, 1979); (see also Metals Tech., 6 (1979) 121-9 (F).

1896. B.H.KEAR, P.R.HOLIDAY and A.R.COX, "Microstructure of Rapidly Solidified IN-100 Powders", Met. Trans., A10 (1979) 191-7 (AQS).

1897. M.J.KELLY and D.W.BULLETT, "A Calculation of the Electronic Structure of the Amorphous $Pd_{80}Si_{20}$ System", J. Phys. C : Solid State Phys., 12 (1979) 2531-41 (SMA).

1898. T.KEMENY, I.VINCZE, B.FOGARASSY and S.ARAJS, "Structure and Crystallization of Fe-B Metallic Glasses", Phys. Rev., B20 (1979) 476-88 (AQS, AB)(also as Report KFKI-78-56, Hungarian Acad. Sci., Budapest (1978), 49pp).

1899. R.KERN and U.GONSER, "Transverse Magnetoresistance of Some Amorphous Metals at 4.2 and 300 K", J. Mag. Mag. Mater., 13 (1979) 74-6 (Mag).

1900. A.J.KERNS, D.E.POLK, R.RAY and B.C.GIESSEN, "Thermal Behaviour of Zr-Cu Metallic Glasses", Mater. Sci. & Engg., 38 (1979) 49-53 (T).

1901. H.R.KHAN and Ch. J.RAUB, "On the Electrical Resistance of Metallic Glass $Pd_{68.5}Cu_{15}Si_{16.5}$", J. Less-Common Metals, 64 (1979) P1-5 (E).

1902. V.P.KHLYNTSEV, L.P.POTAPOV and V.I.KIRIYENKO, "Field-Ion Microscopy Study of Ni-Nb Amorphous Alloy Transition into Crystalline State", Fiz. Met. Metalloved., 47 (1979) 1316-9 (in Russian); Phys. Met. Metallogr., 47, No. 6 (1979) (AB), (see Phys. Abstr., 82(1979) 82158).

1903. M.KIKUCHI, K.FUKAMICHI and T.MASUMOTO, "Invar- and Elinvar-Type Amorphous Fe-Cr-B Alloys with High Corrosion Resistance", J. Japan Inst. Metals, 43 (1979) 885-90 (in Japanese) (M, Mag).

1904. M.KIKUCHI, K.FUKAMICHI and T.MASUMOTO, "The ΔE Effect in Fe-B and Fe-Cr-B Amorphous Invar Alloys", J. Mag. Mag. Mater., 10 (1979) 300-2 (M, Mag).

1905. Y.O.KIM, P.R.STRUTT and H.NOWOTNY, "Laser Melting and Heat Treatment of M2 Tool Steel : A Microstructural Characterization", Met. Trans., A10 (1979) 881-6 (AQS).

1906. W.K.KINNER, "Rapid Quenching : A Mateirals Bonanza", Materials & Energy, 89 (1979) 32-6 (R)(see Met. Abstr., 12(1979) 12-0555).

1907. M.KOBAYASHI, T.HUJISAWA, K.HASHIMOTO and R.YOSOMIYA, "Analysis of Melt Spinning Process of Glass", J. Soc. Mater. Sci. Japan, 28 (1979) 278-82 (in Japanese) (PM)(see Phys. Abstr., 82(1979) 78911).

1908. O.KOHMOTO, K.OHYA, N.YAMAGUCHI, H.FUJISHIMA and T.OJIMA, "Magnetic Properties of Zero Magnetostrictive Amorphous Fe-Co-Si-B Alloys", J. Appl. Phys., 50 (1979) 5054-6 (Mag).

1909. U.KÖSTER, "The Influence of the Specimen Surface on Thermally Induced Structural Changes", Krist. & Technik, 14 (1979) 1369-76 (AB).

1910. E.A.KRAMER, W.L.JOHNSON and C.CLINE, "The Effects of Neutron Irradiation on a Superconducting Metallic Glass", Appl. Phys. Letters, 35 (1979) 815-8 (S).

1911. J.K.KRAUSE, T.C.LONG, D.G.ONN and F.E.LUBORSKY, "Metalloid Effects in the Low Temperature Specific Heat of Transition Metal Metallic Glasses", J. Appl. Phys., 50 (1979) 7665-7 (T).

1912. J.KREMPASKÝ and V.MACKOVA, "The Wiedmann-Franz Law in Semiconductor and Metallic Glasses", Acta Phys. Slov., 29 (1979) 56-66 (E)(see Met. Abstr., 12(1979) 32-0908).

1913. J.KREMPASKÝ, J.VAJDA, A.ZENTKO and P.DUHAJ, "Peculiarities in the Temperature Dependence of Electrical Resistivity of Metallic Glasses", Phys. Stat. Sol., (a) 52 (1979) 387-96 (E).

1914. D.J.KRENITSKY and D.G.AST, "Temperature Dependence of the Flow Stress and Ductility of Annealed and Unannealed Amorphous $Fe_{40}Ni_{40}P_{14}B_6$", J. Mater. Sci., 14 (1979) 275-86 (M).

1915. R.KRISHNAN, SHIVA PRASAD and K.BRANSKA, "Magnetic Characterization of METGLAS 2826 MB Ribbon", J. Appl. Phys., 50 (1979) 7639-41 (Mag).

1916. R.KRISHNAN, M.TESSIER and J.SZTERN, "Magnetic Properties of Amorphous $Fe_{38}Ni_{38}B_{24}$ Alloy", Phys. Stat. Sol., (a) 54 (1979) K77-9 (Mag).

1917. H.KRONMÜLLER, "Micromagnetism in Amorphous Alloys", IEEE Trans. Mag., MAG-15 (1979) 1218-25 (Mag).

1918. H.KRONMÜLLER, M.FÄHNLE, M.DOMANN, H.GRIMM and B.GRÖGER, "Magnetic Properties of Amorphous Ferromagnetic Alloys", J. Mag. Mag. Mater., 13 (1979) 53-70 (Mag).

1919. F.A.KUHNAST, F.MACHIZAUD. R.VENGELSTI and J.FLECHON, "Study of the Evolution of the Morphological, Structural and Physical Properties of the Alloys $Ni_{66}B_{34}$ Deposited in Thin Films or in Powder", J. Microsc. Spectrosc. Electron., 4 (1979) 553-64 (in French) (AB, E).

1920. T.KULIK, J.BASZKIEWICZ, M.KAMINSKI, J.LATUSZKIEWICZ and H.MATYJA, "The Electrochemical Corrosion of Amorphous Ni_{36}-$Fe_{32}Cr_{14}P_{12}B_6$ Alloy (Metglas 2826A)", Corrosion Sci., 19 (1979) 1001-6 (C).

1921. T.KULIK, J.LATUSZKIEWICZ and H.MATYJA, "Crystallization of Metallic Glass $Ni_{36}Fe_{32}Cr_{14}P_{12}B_6$", Arch. Hutn. (Poland), 24 (1979) 103-16 (AB)(see Phys. Abstr., 82(1979) 77520).

1922. H.U.KÜNZI, K.AGYEMAN and H.-J.GÜNTHERODT, "Internal Friction Peaks in Metallic Glasses", Solid State Commun., 32 (1979) 711-4 (M).

1923. Y.C.KUO, C.F.LI, L.M.MAI and W.M.CHANG, "The Magnetic Properties of Fe-Co Amorphous Ribbon in Low Fields", J. Appl. Phys., 50 (1979) 1656-7 (Mag).

1924. L.D.LAFLEUR, "Mössbauer Study of Stress-Induced Rotation of Magnetization in Amorphous $Fe_{80}B_{20}$", Phys. Rev., B20 (1979) 2581-5 (Mag).

1925. J.C.LASJAUNIAS, A.RAVEX and D.THOULOUZE, "Low Temperature Specific Heat of Glassy and Crystallized $Pd_{0.775}Cu_{0.06}Si_{0.165}$ Alloys", J. Phys. F : Metal Phys., 9 (1979) 803-14 (T).

1926. R.M.LATANISION, J.C.TURN, Jr. and C.R.COMPEAU, "The Corrosion Resistance of Metallic Glasses", in Proc. Conf. Mechanical Behavior of Materials, eds. K.J.Miller and R.F.Smith, (Pergamon Press, Oxford, England, 1979), Vol. 2, pp. 475-83 (C, R)(see Met. Abstr., 13(1980) 35-0042).

1927. D.LAZARUS, "Effect of Pressure on Electrical Resistance of $Pd_{82-x}V_xSi_{18}$ Metallic Glass", Solid State Commun., 32 (1979) 175-7 (E).

1928. G.LE CAER and J.M.DUBOIS, "Evaluation of Hyperfine Parameter Distributions from Overlapped Mössbauer Spectra of Amorphous Alloys", J. Phys. E : Sci. Instrum, 12 (1979) 1083-90 (Mag).

1929. B.G.LEWIS, H.A.DAVIES and K.D.WARD, "The Crystallization and Associated Changes in Ductility of Some Fe-Ni-B Glassy Alloys", Scripta Met., 13 (1979) 313-7 (AB, M).

1930. B.G.LEWIS, I.W.DONALD and H.A.DAVIES, "Formation of Filamentary Metallic Glasses by Ultrarapid Quenching from the Melt", in Solidification and Casting of Metals (The Metals Society, London, 1979) pp. 490-5 (AS).

1931. H.H.LIEBERMANN, "Manufacture of Amorphous Alloy Ribbons", IEEE Trans. Mag., MAG-15 (1979) 1393-7 (QT).

1932. C.H.LIN, J.BEVK and D.TURNBULL, "Effect of Structural Relaxation on the Electrical Resistivity of $Pd_{82-x}V_xSi_{18}$ Amorphous Alloys", Solid State Commun., 29 (1979) 641-4 (E).

1933. C.M.LIU, R.INGALLS, J.E.WHITMORE, K.V.RAO and S.M.BHAGAT, "Mössbauer Effect in Ferromagnetic Metallic Glasses : High Pressure Studies", J. Appl. Phys., 50 (1979) 1577-9 (Mag).

1934. J.D.LIVINGSTON, "Stress and Magnetic Domains in Amorphous Metal Ribbons", Phys. Stat. Sol., (a) 56 (1979) 637-45 (Mag).

1935. F.E.LUBORSKY, "Can a Solid be Amorphous and Yet be Ferromagnetic?", Naval. Res. Rev., 32, No. 1 (1979) 4-10 (Mag, R) (see Chem. Abstr., 92(1980) 50983t).

1936. F.E.LUBORSKY, J.J.BECKER, J.L.WALTER and H.H.LIEBERMANN, "Formation and Magnetic Properties of Fe-B-Si Amorphous Alloys", IEEE Trans. Mag., MAG-15 (1979) 1146-9 (Mag).

1937. F.E.LUBORSKY, P.J.FLANDERS, H.H.LIEBERMANN and J.L.WALTER, "Effect of Metalloid on Magnetostriction of Fe-B-C and Fe-B-Si Amorphous Alloys", IEEE Trans. Mag., MAG-15 (1979) 1961-2 (Mag).

1938. F.E.LUBORSKY, T.KEMÈNY and I.VINCZE, "Electrical Resistivity of Glassy $(Fe_xNi_{100-x})_{80}B_{20}$ (x=30, 40, 50, 60, 70) Alloys at Elevated Temperatures", J. Appl. Phys., 50 (1979) 1654-5 (E).

1939. F.E.LUBORSKY, J.L.WALTER and H.H.LIEBERMANN, "Engineering Magnetic Properties of Fe-Ni-B Amorphous Alloys", IEEE Trans. Mag., MAG-15 (1979) 909-11 (Mag).

1940. H.L.LUO and J.G.HUBER, "Superconductivity in Th-Zr Alloys", J. Less-Common Metals, 65 (1979) 13-7 (S).

1941. F.MACHIZAUD, F.A.KUHNAST and J.FLECHON, "Electronic Transport and Determination of Apparent Activation Energy for Crystallization of Amorphous Alloys $Ni_{71}B_{29}$ and $Ni_{66}B_{34}$", Thin Solid Films, 61 (1979) 327-34 (in French) (AB)(see Phys. Abstr., 83 (1980) 6807).

1942. J.U.MADSEN, "Evidence for Phase Separation Prior to Crystallization in Annealed Fe-Ni Metallic Glasses", Scand. J. Metallurgy, 8 (1979) 27-30 (AB).

1943. J.U.MADSEN and R.M.J.COTTERILL, "On the Size Statistics of Bernal Holes", Phys. Stat. Sol., (a) 56 (1979) K5-10 (SMA).

1944. A.K.MAJUMDAR, "Magnetic Properties of Amorphous Ferromagnet $Fe_{78}Mo_2B_{20}$", Solid State Commun., 29 (1979) 85-8 (Mag).

1945. Y.MAKINO, "Amorphous Alloys for Head Materials", Kinzoku, 49 No. 9 (1979) 8-14 (in Japanese) (A, R).

1946. R.MALMHÄLL, G.BÄCKSTRÖM, S.M.BHAGAT and K.V.RAO, "Hall Resistivity of a Two-Phase Magnetic Material - Crystallized Metglas 2826", J. Phys. F: Metal Phys., 9 (1979) 317-25 (Mag).

1947. R.MALMHÄLL, G.BÄCKSTRÖM, K.V.RAO and T.EGAMI, "Compositional Short-Range Order Effects in the Hall Resistivity of Amorphous $Fe_{27}Ni_{53}P_{14}B_6$", J. Appl. Phys., 50 (1979) 7656-8 (Mag).

1948. R.MALMHÄLL, S.M.BHAGAT, K.V.RAO and G.BÄCKSTRÖM, "Transport Properties of Amorphous Ferromagnets - Hall and Ohmic Resistivities", Phys. Stat. Sol., (a) 53 (1979) 641-51 (E, Mag).

1949. A.P.MALOZEMOFF, P.H.CHANG and M.GRIMSDITCH, "Brillouin Scattering from Rapidly Quenched Ferromagnetic Metallic Glasses", J. Appl. Phys., 50 (1979) 5896-8 (AQS).

1950. A.P.MALOZEMOFF, W.FERNENGEL and A.BRUNSCH, "Observation and Theory of Strong Stripe Domains in Amorphous Sputtered FeB Films", J. Mag. Mag. Mater., 12 (1979) 201-14 (Sp; Mag).

1951. A.P.MALOZEMOFF, M.GRIMSDITCH, J.ABOAF and A.BRUNSCH, "Brillouin Scattering Studies of Polycrystalline and Amorphous Sputtered Films of $Fe_{1-x}B_x$ and $Co_{1-x}B_x$", J. Appl. Phys., 50 (1979) 5885-95 (AQS).

1952. G.MARCHAL, Ph.MANGIN, M.PIECUCH and Chr.JANOT, "Some Comments about Crystallization of Fe-Si Amorphous Films", Phys. Stat. Sol., (a) 56 (1979) K29-31 (AB).

1953. M.A.MARCUS, "Electrical Resistometric Detection of Relaxation in an Amorphous Pd-Si-Sb Alloy", Acta Met., 27 (1979) 879-91 (E).

1954. M.A.MARCUS, "Mechanical Properties of Mostly Crystalline, Rapid-Quenched Pd-Si-Sb", Acta Met., 27 (1979) 893-902 (M).

1955. M.A.MARCUS, "Phase Separation and Crystallization in Amorphous Pd-Si-Sb", J. Non-Cryst. Solids, 30 (1979) 317-35 (AB).

1956. R.E.MARINGER and C.E.MOBLEY, "Direct Casting of Wire, Filament and Fiber", Wire J., 12 (1979) 70-4 (QT).

1957. N.D.MARKELOV and E.Z.SPEKTOR, "Characteristics of Structural Factors of Molten and Amorphous Alloys", Doklady Adak. Nauk SSSR, 248 (1979) 851-4 (in Russian) (AQS)(see Chem. Abstr., 92(1980) 80217p).

1958. Ž.MAROHNIĆ, K.SAUB, E.BABIĆ and J.IVKOV, "Resistivity Variation near the Critical Temperature in Some Ni-Fe Base Amorphous Alloys", Solid State Commun., 30 (1979) 651-5 (E).

1959. M.MATSUURA, "Crystallization Kinetics of Amorphous Fe-B Alloys by DTA", Solid State Commun., 30 (1979) 231-3 (AB).

1960. A.MATSUZAKI, S.NANAO and H.INO, "Structure and Bonding Nature of Pd-Si Amorphous Alloys", J. Physique, 40 (1979) C2-104-6 (AQS).

1961. T.MATSUZAWA, K.OKAMURA, T.SHISHIDO and S.YAJIMA, "Syntheses of Amorphous and 'Orthorhombic' GIG", J. Physique, 40 (1979) C2-149-50 (AS, MCS, AB).

1962. T.R.McGUIRE and R.J.GAMBINO, "Magnetic and Transport Properties of Rare-Earth Au and Cu Amorphous Alloys", J. Appl. Phys., 50 (1979) 7653-5 (Sp; Mag, T).

1963. J.MEGUSAR, A.S.ARGON and N.J.GRANT, "Plastic Flow and Fracture in $Pd_{80}Si_{20}$ near T_g", Mater. Sci. & Engg., 38 (1979) 63-72 (M).

1964. F.MEYER, J.H.J M.BUSTER, B.G.BAGLEY and D.E.ASPNES, "Optical Properties of the Metallic Glass $Pd_{0.775}Cu_{0.06}Si_{0.165}$ Over the Energy Range of 0.67 to 5.6 eV", J. Non-Cryst. Solids, 34 (1979) 441-4 (PP).

1965. S.A.MILLER and R.J.MURPHY, "A Gas-Water Atomization Process for Producing Amorphous Powders", Scripta Met., 13 (1979) 673-6 (QT).

1966. G.v.MINNIGERODE and K.SAMWER, "Electrons in Metallic Glasses", Revue de Chimie Minerale, 16 (1979) 370-80 (R).

1967. I.S.MIROSHNICHENKO, "Formation of Metastable Phases in a Rapidly Cooled Melt", in Growth of Crystals, ed. A.A.Chernov (Consultants Bureau, New York, 1979), Vol. II, pp. 344-54 (MCS).

1968. M.A.MITCHELL, J.R.CULLEN, R.ABBUNDI, A.CLARK and H.SAVAGE, "Magnetoelastic Effects in $Fe_{71}Co_9B_{20}$ Glassy Ribbons", J. Appl. Phys., 50 (1979) 1627-9 (M, Mag).

1969. M.MITERA, T.MASUMOTO and N.S.KAZAMA, "Effect of Silicon Addition on the Magnetic Properties of Fe-B-C Amorphous Alloys", J. Appl. Phys., 50 (1979) 7609-11 (Mag).

1970. K.MIYAZAWA and J.SZEKELY, "A Mathematical Model of the Splat Cooling Process Using the Piston and Anvil Technique", Met. Trans., B10 (1979) 349-58 (PM).

1971. T.MIZOGUCHI, "Magnetic Properties of Amorphous Materials", Kinzoku, 49, No. 9 (1979) 2-7 (in Japanese) (R, Mag).

1972. T.MIZOGUCHI and G.S.CARGILL III, "Magnetic Anisotropy from Dipolar Interactions in Amorphous Ferrimagnetic Alloys", J. Appl. Phys., 50 (1979) 3570-82 (Mag).

1973. A.MOGRO-COMPERO and J.L.WALTER, "Resistivity of Amorphous Fe-B Alloys : Measurements and Implications", Phys. Rev., B20 (1979) 5030-4 (E).

1974. B.V.MOLOTILOV and Yu.A.GRATSIANOV, "Procedures for the Production and Prospects of Using Amorphous Metallic Materials for Developing Precision Alloys", Pretsizion Splavy, No. 5 (1979) 5-12 (in Russian) (A)(see Chem. Abstr., 91(1979) 215125g).

1975. H.MORITA, H.FUJIMORI and Y.OBI, "Roll Magnetic Anisotropy in Amorphous Ferromagnet $Fe_{4.7}Co_{70.3}Si_{15}B_{10}$", Japan J. Appl. Phys., 18 (1979) 683-4 (Mag).

1976. H.MORITA, H.FUJIMORI and Y.OBI, "Magnetic Anisotropy of Amorphous $(Fe_{1-x}Co_x)_{78}Si_{10}B_{12}$ Alloys", Appl. Phys., 20 (1979) 125-7 (Mag).

1977. R.C.MORRIS, "Disclination-Dislocation Model of Metallic Glass Structures", J. Appl. Phys., 50 (1979) 3250-7 (SMA).

1978. D.MUSSER, C.L.CHEN and H.S.CHEN, "Mössbauer Study of a Binary Amorphous Magnet : $Fe_{80}P_{20}$", J. Appl. Phys., 50 (1979) 7659-61 (Mag).

1979. D.MUSSER, C.L.CHIEN, F.E.LUBORSKY and J.L.WALTER, "The Influence of Carbon on the Hyperfine Interaction and Curie Temperatures of Amorphous $Fe_{84}B_{16-x}C_x$", J. Appl. Phys., 50 (1979) 1571-3 (Mag).

1980. M.NAGUMO, "Embrittlement of Amorphous Metals", Kinzoku, 49, No. 12 (1979) 20-3 (in Japanese) (R, M).

1981. M.NAKA, K.HASHIMOTO, A.INOUE and T.MASUMOTO, "Corrosion Resistant Amorphous Fe-C Alloys Containing Cr and/or Mo", J. Non-Cryst. Solids, 31 (1979) 347-54 (C).

1982. M.NAKA, K.HASHIMOTO and T.MASUMOTO, "Change in Corrosion Behaviour of Amorphous Fe-P-C Alloys by Alloying with Various Metallic Elements", J. Non-Cryst. Solids, 31 (1979) 355-65 (C).

1983. M.NAKA, K.HASHIMOTO and T.MASUMOTO, "Effect of Addition of Cr and Mo on the Corrosion Behaviour of Amorphous Fe-20B and Ni-20B Alloys", J. Non-Cryst. Solids, 34 (1979) 257-66 (C).

1984. S.NANAO, J.S.SIGIURA, H.INO, Y.MAEDA and H.SAKAI, "A Study of Local Atomic and Electronic Structures of Liquid-Quenched La-Au Alloys by ^{197}Au and ^{57}Fe Spectroscopy", J. Physique, 40 (1979) C2-123-5 (AQS).

1985. M.NAOE, H.YAMAMOTO and S.YAMANAKA, "Magnetic Properties of Amorphous Fe-Si-B Films Prepared by Co-Sputtering", J. Appl. Phys., 50 (1979) 7606-8 (Sp; Mag).

1986. K.NARITA, J.YAMASAKI and H.FUKUNAGA, "Saturation Magnetostriction and Its Annealing Behavior of $Fe_{100-x}B_x$ and Co_{100-x}-B_x Amorphous Alloys", J. Appl. Phys., 50 (1979) 7591-3 (Mag).

1987. K.NASSAU, M.GRASSO and A.M.GLASS, "Quenched Glasses in the Systems of Li_2O with Al_2O_3, Ga_2O_3 and Bi_2O_3", J. Non-Cryst. Solids, 34 (1979) 425-36 (AQS, T, MCS).

1988. K.NASSAU, C.A.WANG and M.GRASSO, "Quenched Metastable Glassy and Crystalline Phases in the System Li-Na-K-Metatantalate", J. Amer. Ceram. Soc., 62 (1979) 74-9 (AS, MCS).

1989. P.K.K.NAYAR, "Preferred Orientation in Splat-Quenched Zinc, Cadmium and Bismuth Foils", Scripta Met., 13 (1979) 1115-6 (AQS).

1990. T.J.NEGRAN and A.M.GLASS, "A Study of the Crystallization Kinetics of Roller Quenched $LiNbO_3$ Glass Using Trivalent Chromium Ion Luminescence as a Probe", Phys. & Chem. Glasses, 20 (1979) 140-1 (AB).

1991. D.A.NEW, S.ARAJS, F.E.LUBORSKY, T.KEMÈNY and I.VINCZE, "Electrical Resistivity of Glassy $(Fe_xNi_{100-x})_{80}B_{20}$ (x=30, 40, 50, 60, 70) Alloys at Elevated Temperatures", J. Appl. Phys., 50 (1979) 1654-5 (E).

1992. Y.OBI, "The Stress Effect on Magnetic Domain Structure of Fe-Co Amorphous Alloys", Appl. Phys., 18 (1979) 119-21 (Mag).

1993. Y.OBI, H.MORITA and H.FUJIMORI, "Exchange Anisotropy in Ni-Mn-Si-B Amorphous Alloys", Phys. Stat. Sol., (a) 55 (1979) K67-70 (Mag).

1994. P.OELHAFEN, E.HAUSER, H.-J.GÜNTHERODT and K.H.BENNEMANN, "New Type of d-Band Metal Alloys : The Valence Band Structure of the Metallic Glasses Pd-Zr and Cu-Zr", Phys. Rev. Letters, 43 (1979) 1134-7 (AQS).

1995. P.OELHAFEN, M.LIARD, H.-J.GÜNTHERODT, K.BERRESHEIM and H.D.POLASCHEGG, "Photoemission (XPS, UPS) Studies of Pd-Si Metallic Glasses", Solid State Commun., 30 (1979) 641-4 (AQS).

1996. R.C.O'HANDLEY, C.P.CHOU and N.DE CRISTOFARO, "High Induction Low Loss Metallic Glasses", J. Appl. Phys., 50 (1979) 3603-7 (Mag).

1997. R.C.O'HANDLEY and M.C.NARASIMHAN, "Low-Field Magnetic Properties of Wide Fe-Ni Metallic Glass Strips", IEEE Trans. Mag., MAG-15 (1979) 970-2 (Mag).

1998. R.C.O'HANDLEY, M.C.NARASIMHAN and M.O.SULLIVAN, "Magnetostriction of $Fe_{100-x}B_x$ Glasses", J. Appl. Phys., 50 (1979) 1633-5 (Mag).

1999. S.OHNUMA and T.MASUMOTO, "Zero Magnetstriction and Low-Field Magnetic Properties of Co-Fe Base Amorphous Alloys Containing V, Nb, Ta, Cr, Mo, W, Mn or Ni", J. Appl. Phys., 50 (1979) 7597-9 (Mag).

2000. T.OKADA, S.MATSUSHITA and Y.SAKURAI, "Amorphous Fe-Co-Si-B Sputtered Thin Films", Japan J. Appl. Phys., 18 (1979) 2305-6 (Sp; Mag).

2001. K.ONISHI, H.TONOMURA and Y.SAKURAI, "Measurement of Magnetic Potential Distribution and Wall Velocity in Amorphous Films", J. Appl. Phys., 50 (1979) 7624-6 (Sp; Mag).

2002. H.ONODERA, H.YAMAMOTO and H.WATANABE, "Temperature Dependence of Magnetic Anisotropy in Amorphous Fe-B Alloys", J. Physique, 40 (1979) C2-142-3 (Mag).

2003. J.OREHOTSKY, "Isothermal Crystallization Kinetics of Amorphous $Fe_{80}(B_{1-x}P_x)_{20}$ Alloys", J. Appl. Phys., 50 (1979) 7612-4 (AB).

2004. K.OSAMURA, K.SHIBUE, P.H.SHINGU and Y.MURAKAMI, "Small-Angle X-Ray Scattering Study on the Structure and Crystallization of Amorphous Fe-P-C Alloy", J. Mater. Sci., 14 (1979) 945-52 (AQS, AB).

2005. R.OSHIMA, F.E.FUJITA, K.FUKAMICHI and T.MASUMOTO, "Mössbauer Spectroscopy of Amorphous Fe-B Alloys", J. Physique, 40 (1979) C2-132-4 (Mag).

2006. K.J.OVERSHOTT, "Amorphous Ribbon Materials and Their Possibilities", Electron Power, 25 (1979) 347-50 (A)(see Met. Abstr., 13(1980) 33-0432).

2007. K.J.OVERSHOTT and C.D.GRAHAM, Jr., "Power Loss Measurements in Amorphous Alloys", J. Appl. Phys., 50 (1979) 1598-1600 (Mag).

2008. J.P.PATTERSON and D.R.H.JONES, "The Effect of Homogeneous Deformation on the Crystallization of a Metallic Glass", Scripta Met., 13 (1979) 947-9 (AB).

2009. J.P.PATTERSON and D.R.H.JONES, "Viscous Flow and Forming of a Metallic Glass", in Proc. Conf. Mechanical Behavior of Materials, eds. K.J.Miller and R.F.Smith (Pergamon Press, Oxford, England, 1979), Vol. 2, pp. 547-56 (M, F)(see Met. Abstr., 13(1980) 31-0187).

2010. D.I.PAUL and L.M.PECORA, "Temperature Dependence of the Magnetization for a Random-Uniaxial-Anisotropy Amorphous Ferromagnetic Material ", Phys. Rev., B19 (1979) 4608-13 (Mag).

2011. S.J.PICKART, Y.P.SHARMA and J.P.DE NEUVILLE, "Structural Analysis of Amorphous $Ge_{0.5}Te_{0.5}$ by Neutron Diffraction", J. Non-Cryst. Solids, 34 (1979) 183-90 (AQS).

2012. J.T.PRATER and T.EGAMI, "Annealing Effects in Ni-Based Amorphous Alloys with Dilute Concentrations of Fe", J. Appl. Phys., 50 (1979) 1706-8 (Mag).

2013. B.PREDEL and K.HÜLSE, "Metastable Crystalline and Glassy Phases in the System Mg-Ga", J. Less-Common Metals, 63 (1979) 45-56 (in German) (TSSE, MCS, AS).

2014. B.PREDEL and K.HÜLSE."Thermodynamic Properties of Metastable Phases in the System Mg-Ga", J. Less-Common Metals, 63 (1979) 245-56 (in German) (T).

2015. G.QUINTANA, H.SIRKIN, M.ROSEN, D.H.KURLAT and E.FRANK, "Mössbauer Spectra for a Splat-Cooled SnPb Eutectic Alloy", Phil. Mag., B39 (1979) 457-9 (AQS).

2016. P.RAMACHANDRARAO, K.LAL, A.SINGHDEO and K.CHATTOPADHYAY, "Quenching from the Mushy State - A New Technique for the Production of Metastable Phases", Mater. Sci. & Engg., 41 (1979) 259-64 (QT, MCS).

2017. K.V.RAO, H.GUDMUNDSOON, H.U.ASTROM and H.S.CHEN, "Influence of Metal-Metalloid Elements on the Electrical Resistance of Amorphous Materials", J. Appl. Phys., 50 (1979) 1592-4 (E).

2018. O.RAPP, J.E.GRINDBORG and K.V.RAO, "Low Temperature Electrical Resistance of Disordered $(Fe_{1-x}Ni_x)_{75}P_{16}B_6Al_3$ Alloys", J. Low-Temp. Phys., 35 (1979) 89-102 (E).

2019. J.J.RHYNE, J.W.LYNN, F.E.LUBORSKY and J.L.WALTER, "Spin Excitations in Amorphous Transition Metal Boron Glasses", J. Appl. Phys., 50 (1979) 1583-5 (Mag).

2020. R.RICHTER, M.WOLF and F.GOEDSCHE, "On the Spin-Disorder Resistivity of Amorphous Ferromagnets", Phys. Stat. Sol., (b) 95 (1979) 473-82 (E).

2021. J.D.RILEY, L.LEY, J.AZOULAY and K.TERAKURA, "Partial Densities of States in Amorphous $Pd_{0.81}Si_{0.19}$", Phys. Rev., B20 (1979) 776-83 (AQS).

2022. J.M.RIVEIRO and C.AROCA, "Effect of the Anisotropy Constant on the Strip Domains of Metglas 2826 and $Co_{85}P_{15}$ Amorphous Materilas", J. Mag. Mag. Mater., 13 (1979) 289-93 (Mag).

2023. J.RIVORY and B.BOUCHET, "Electrical and Optical Properties of CoP Metallic Glasses", J. Phys. F : Metal Phys., 9 (1979) 327-35 (E, PP).

2024. D.ROGALLA and G.TRIPPEL, "Amorphous Soft Magnetic $Fe_{80}B_{20}$ Film with Ni Addition", IBM Tech. Disclosure Bull., 21 (1979) 3226 (Mag)(see Phys. Abstr., 82(1979) 82836).

2025. M.SAKATA, N.COWLAM and H.A.DAVIES, "Neutron Diffraction Measurement of the Structure Factor of a CuTi Metallic Glass", J. Phys. F : Metal Phys., 9 (1979) L235-40 (AQS).

2026. I.R.SARE and R.W.K.HONEYCOMBE, "Microstructural Effects of Splat Cooling a High-Speed Steel", Metal Sci., 13 (1979) 269-79 (AQS).

2027. G.V.S.SASTRY and C.SURYANARAYANA, "An Electron Diffraction Study of a Metastable Al_2Pd Phase", J. Less-Common Metals, 63 (1979) P 89-91 (MCS).

2028. A.S.SCHAAFSMA, H.SNIJDERS, F.VAN DER WOUDE, J.W.DRIJVER and S.RADELAAR, "Amorphous to Crystalline Transformation of $Fe_{80}B_{20}$", Phys. Rev., B20 (1979) 4423-30 (AB, Mag).

2029. A.SCHLEGEL, P.WACHTER, K.P.ACKERMANN, M.LIARD and H.-J.GÜNTHERODT, "Optical Properties of Metallic Glasses", Solid State Commun., 31 (1979) 373-6 (PP).

2030. K.W.SCHLENK, "Amorphous Metals - Metals with the Structure of Glass", Feinwerktech. & Messtech., 87, No. 1 (1979) 44-5 (in German) (R)(see Phys. Abstr., 82(1979) 51813).

2031. J.SCHNEIDER and W.POMPE, "Metallic Glasses - Preparation, Structure and Physical Properties", Neue Huette, 24 (1979) 379-83 (in German) (R)(see Chem Abstr., 92(1980) 62925p).

2032. J.SCHNEIDER, B.SPRINGMANN, K.ZAVETA, G.WENZEL and E.KÖHLER, "Magnetic After-Effect in Amorphous Fe-Co-Based Alloys", Phys. Stat. Sol., (a) 52 (1979) K25-9 (Mag).

2033. J.SCHNEIDER, K.ZAVETA, A.HANDSTEIN, F.ZOUNOVA and V.NEKVASIL, "Temperature Dependence of the Magnetization in Amorphous $(Fe_{0.04}Ni_{0.96})_{80}P_{10}B_{10}$", Phys. Stat. Sol., (a) 56 (1979) K141-4 (Mag).

2034. P.J.SCHURER and A.H.MORRISH, "Mössbauer Study of Amorphous $Fe_{78}B_{12}Si_{10}$", Phys. Rev., B20 (1979) 4660-4 (AQS).

2035. P.J.SCHURER and A.H.MORRISH, "Long- and Short-Range Exchange Interactions in Ferromagnetic Amorphous Iron Alloys", Solid State Commun., 30 (1979) 21-4 (Mag).

2036. M.G.SCOTT and T.WATANABE, "Crystallization of Some Metallic Glasses", in Phase Transformations (Institution of Metallurgists, London, 1979), Part III, pp. 36-8 (AB)(see Phys. Abstr., 83 (1980) 6367).

2037. S.SEKI, "Amorphous State and Glassy State", Kinzoku, 49, No.9 (1979) 29-32 (in Japanese) (AQS, R).

2038. J.E.SHELBY, "Thermal Expansion of Amorphous Metals", J. Non-Cryst. Solids, 34 (1979) 111-9 (T).

2039. T.SHI, L.ZHENG and N.CHEN, "Prediction on Binary Amorphous Alloy Formation with Band-Parametric Diagram", Acta Met. Sinica, 15 (1979) 94-7 (in Chinese) (PM)(see Phys. Abstr., 83(1980) 11587).

2040. T.SHIGEMATSU, T.SHINJO, Y.BANDO and T.TAKADA, "Formation of Metastable Phase in Vapor-Deposited Iron Antimonide Films", Bull. Inst. Chem. Res., Kyoto Univ., 57z (1979) 310-7 (MCS, AS, AB)(see Chem. Abstr., 92(1980) 87064v).

2041. K.SHIIKI, T.WATANABE and M.KUDO, "Magnetic Instability of $(Fe_{0.06}Co_{0.94})$-B Amorphous Alloy System", J. Appl. Phys., 50 (1979) 5419-24 (Mag).

2042. Y.SHIMADA and H.KOJIMA, "Magnetic Properties and Thermal Stability of Fe-B Amorphous Films", J. Appl. Phys., 50 (1979) 1541-3 (Sp; Mag, T).

2043. Y.SHIMADA and H.KOJIMA, "Amorphous Ferromagnetic Films Produced by Sputtering", Bull. Japan Inst. Metals, 18 (1979) 413-8 (in Japanese) (Sp; R).

2044. P.H.SHINGU, K.SHIMOMURA and R.OZAKI, "Fabrictaion of Amorphous Alloy Sheet by Gas Melt-Spraying Method", Trans. Japan Inst. Metals, 20 (1979) 33-5 (QT).

2045. K.SHIRAE, "Compatibility of DC and AC Magnetic Properties of Amorphous Magnetic Materials", J. Appl. Phys., 50 (1979) 7618-20 (Mag).

2046. K.SHIRAKAWA and T.MASUMOTO, "The Strain Gauge Factor and Electrical Properties in Non-Magnetic Amorphous Cu-Ag-P Alloys", J. Japan Inst. Metal, 43 (1979) 634-6 (in Japanese) (E).

2047. T.SHISHIDO, "The Crystallization of Gd_2O_3-Al_2O_3 Glasses", J. Mater. Sci., 14 (1979) 823-30 (AB).

2048. T.SHISHIDO and K.OKAMURA, "Amorphous Complex Oxides Obtained by Rapid Quenching", Bull. Japan Inst. Metals, 18 (1979) 740-7 (AQS, R).

2049. SHIVA PRASAD, R.KRISHNAN, G.SURAN and J.SZTERN, "Inhomogeneities and FMR Spectra in Some D.C.Sputtered Gd-Co Films", J. Appl. Phys., 50 (1979) 1623-5 (Sp; AQS).

2050. M.SOSTARICH and W.GOLENIA, "Electrical Resistivity of Metglas 2826A $(Fe_{32}Ni_{36}Cr_{14}P_{12}B_6)$ During Crystallization", Solid State Commun., 31 (1979) 339-42 (E, AB).

2051. F.H.M.SPIT, J.W.DRIJVER and S.RADELAAR, "Hydrogen Sorption in Amorphous Ni(Zr, Ti) Alloys", Z. PhysikalischeChemie, 116 (1979) 225-32 (C).

2052. S.STEEB and P.LAMPARTER, "Investigation of Metals and Amorphous Substances by means of Neutron Diffraction", Z. Metallkde., 70 (1979) 660-8 (R).

2053. L.STEINBERG and S.ANDERSSON, "Electron Microscope Studies on a Quenched Fe-Mo Alloy", J. Solid State Chem., 28 (1979) 269-77 (AQS).

2054. W.M.STOBBS and D.J.SMITH, "High Resolution Imaging of Amorphous Materials", Nature, 281 (1979) 54-5 (AQS).

2055. W.M.STOBBS and J.V.WOOD, "A Rapidly Solidified Marmem Alloy", Acta Met., 27 (1979) 575-84 (AQS).

2056. B.STRITZKER and H.L.LUO, "Superconductivity in Pd-Si-H(D) Alloys", Solid State Commun., 29 (1979) 811-4 (S).

2057. M.STUBICAR, "Microhardness Characterization of Stability of Fe-Ni-Base Metallic Glasses", J. Mater. Sci., 14 (1979) 1245-8 (T, M).

2058. G.SURAN and R.J.GAMBINO, "Spin Wave Spectra of Non-Magnetostrictive Amorphous Alloy Films", J. Appl. Phys., 50 (1979) 7671-3 (Mag).

2059. Y.SYONO and A.INO, "Structural and Magnetic Properties of Amorphous $BaO-Fe_2O_3-B_2O_3$ System", J. Physique, 40 (1979) C2-115-7 (Mag).

2060. Y.SYONO, A.ITO and O.HORIE, "Mössbauer Study on the Structural and Magnetic Properties of Amorphous $BaO-Fe_2O_3-B_2O_3$ System", J. Phys. Soc. Japan, 46 (1979) 793-801 (AQS, Mag).

2061. D.SZYMANSKI, J.C.BARRICK and B.C.GIESSEN, "The Crystal Structure of a Metastable Binary Phase Related to β-Ga: β'-Ga(In)", J. Solid State Chem., 30 (1979) 55-9 (MCS).

2062. L.TAKÁCS, "Electronic Structure of Transition Metal-Metalloid Crystalline and Glassy Alloys", Phys. Stat. Sol., (a) 56 (1979) 371-7 (AQS).

2063. M.TAKAHASHI and M.KOSHIMURA, "Mössbauer Effect of Amorphous Fe-P Alloys", Japan. J. Appl. Phys., 18 (1979) 685-6 (AQS).

2064. M.TAKAHASHI, M.KOSHIMURA and T.SUZUKI, "Mössbauer Effect of Amorphous Fe-B Alloys", J. Physique, 40 (1979) C2-144-6 (AQS).

2065. M.TAKAHASHI and T.MIYAZAKI, "Magnetic Anisotropy in an Amorphous $Fe_{80}P_{13}C_7$ Alloy", Japan. J. Appl. Phys., 18 (1979) 743-52(Mag).

2066. M.TAKAHASHI, T.MIYAZAKI and A.WATANABE, "The Change in Static Magnetic Properties with Isothermal Annealing in Amorphous Ferromagnetic Alloys", J. Japan Inst. Metals, 43 (1979) 339-47 (in Japanese) (Mag).

2067. M.TAKAHASHI, T.MIYAZAKI, T.WATANABE and S.ISHIO, "Magnetic Properties of Rapidly Quenched Fe-Al Alloys", Japan. J. Appl. Phys., 18 (1979) 2325-6 (Mag).

2068. S.TAKAYAMA, "Drawing of $Pd_{77.5}Cu_6Si_{16.5}$ Metallic Glass Wires", Mater. Sci. & Engg., 38 (1979) 41-8 (M).

2069. S.TAKAYAMA, "Serrated Flow in Metallic Glasses", Scripta Met., 13 (1979) 463-7 (M).

2070. S.TAKAYAMA and T.OI, "The Effects of Processing Conditions on Magnetic Properties of Amorphous Alloys", J. Appl. Phys., 50 (1979) 1595-7 (Mag).

2071. S.TAKAYAMA and T.OI, "The Analysis of Casting Conditions of Amorphous Alloys", J. Appl. Phys., 50 (1979) 4962-5 (PM).

2072. I.M.TANG, "Critical Temperature of an Amorphous Ferromagnet", Phys. Letters, A74 (1979) 127-8 (Mag).

2073. S.TANIGAWA, K.HINODE, R.NAGAI, K.KANBE, M.DOYAMA and N.SHIOTANI, "Positron Annihilation in Amorphous Alloys", Phys. Stat. Sol., (a) 51 (1979) 249-55 (AQS).

2074. L.E.TANNER and R.RAY, "Metallic Glass Formation and Properties in Zr and Ti Alloyed with Be -I. The Binary Zr-Be and Ti-Be Systems", Acta Met., 27 (1979) 1727-47 (AQS, T, M, AB).

2075. A.I.TAUB and F.SPAEPEN, "Isoconfiguration Flow of Amorphous $Pd_{82}Si_{18}$", Scripta Met., 13 (1979) 195-8 (M).

2076. A.I.TAUB and F.SPAEPEN, "Enhancement of Structural Relaxation in Amorphous PdSi Following Stress Reduction", Scripta Met., 13 (1979) 883-6 (M).

2077. C.V.THOMPSON and F.SPAEPEN, "On the Approximation of the Free Energy Change on Crystallization", Acta Met., 27 (1979) 1855-9 (T).

2078. S.K.TIWARI, K.CHATTOPADHYAY, C.SURYANARAYANA and T.R.ANANTHARAMAN, "Decomposition Studies of a Liquisol-Quenched Al-33 at.% Mg Alloy", Metallography, 12 (1979) 73-86 (AB, TSSE).

2079. V.I.TKACH, V.F.BASHEV, A.B.LYSENKO and I.S.MIROSHNICHENKO, "Formation of Metastable Samarium Modification during Quenching from Melt", Fiz. Met. Metalloved., 48 (1979) 213-5 (in Russian); Phys. Met. Metallogr., 48, No. 1 (1979) (MCS)(see Phys. Abstr., 82 (1979) 91636).

2080. K.TOGANO, "Current Researches on Amorphous Superconductors", Bull. Japan Inst. Metals, 18 (1979) 675-83 (in Japanese) (R, S).

2081. E.TÖRÖK and G.HAUSCH, "Search for Invar-Like Magnetoelastic Behaviour in Fe-Ni-Based Metallic Glasses", J. Mag. Mag. Mater., 10 (1979) 303-6 (M, Mag).

2082. P.K.TSENG, S.Y.CHUANG and H.S.CHEN, "Mössbauer Effect in an Aged Metallic Glass", J. Appl. Phys., 50 (1979) 4292-4 (AB).

2083. S.TSUNASHIMA, S.MITSUYA and S.UCHIYAMA, "Magnetic and Mageto-elastic Properties of Amorphous Fe-Si Films", Japan. J. Appl. Phys., 18 (1979) 1645-6 (Mag).

2084. N.TSUYA and K.I.ARAI, "Magnetostriction of Ribbon Form Amorphous and Crystalline Ferromagnetic Alloys", J. Appl. Phys., 50 (1979) 1658-63 (Mag).

2085. N.TSUYA and K.I.ARAI, "Rapidly Quenched Ribbon-Form Silicon and Silicon-Iron", Japan. J. Appl. Phys., 18 (1979) 207-8 (QT).

2086. N.TSUYA and K.I.ARAI, "Amorphous Ribbons of $LiNbO_3$, $PbTiO_3$ and $(PbTiO_3)_{0.9}(B_2O_3)_{0.1}$", Japan. J. Appl. Phys., 18 (1979) 461-3 (AS).

2087. N.TSUYA, K.I.ARAI and K.OHMORI, "Ribbon-Form Sendust Alloy", IEEE Trans. Mag., MAG-15 (1979) 1149-53 (QT, Mag).

2088. S.TYAGI and A.E.LORD. Jr., "Elinvar-Type Behaviour in the Alloy $Fe_{78}Mo_2B_{20}$", Appl. Phys. Letters, 34 (1979) 533-4 (M, Mag).

2089. S.TYAGI and A.E.LORD, Jr., "Internal Friction and Shear Modulus of Certain Metallic Glasses", J. Non-Cryst. Solids, 30 (1979) 273-83 (M).

2090. M.UENO, M.MISAWA and K.SUZUKI, "T-O-F Pulsed Neutron Total Scattering Studies of V_2O_5-P_2O_5 and PbO-GeO_2 Glasses Prepared by Rapid Quenching from Their Melts", Res.Rep. Lab. Nucl. Sci. Tohoku Univ., No. 12 (1979) 207-11 (AQS).

2091. G.VAN TENDELOO, J.VAN LANDUYT and S.AMELINCKX, "Radiation Ordering in Quenched Alloys Observed 'in-situ' in High-Voltage Electron Microscope", Radiation Effects, 41 (1979) 179-84 (AQS).

2092. R.D.VELITRI, E.M.BREINAN, G.P.McCARTHY and F.S.GALASSO, "Forming Rapidly Cooled Materials Using a Laser Beam", J. Mater. Sci., 14 (1979) 3000-2 (QT).

2093. J.A.VAN VECHTEN, R.J.GAMBINO and J.J.CUOMO, "Encapsulation of Radioactive Noble Gas Waste in Amorphous Alloy", IBM J. Res. Develop., 23 (1979) 278-85 (A)(see Met. Abstr., 12(1979) 16-0404).

2094. N.I.VARICH and A.N.PETRUNINA, "Some Peculiarities in Crystallization of Rapidly Cooled Peritectic Ni-Ru Alloys", Izv. Akad. Nauk. SSSR, Met., No. 2 (1979) 112 (in Russian) (TSSE, MCS)(see Phys. Abstr., 83(1980) 20922).

2095. I.VINCZE, D.S.BOUDREAUX and M.TEGZE, "Short-Range Orders in Fe-B Metallic Glass Alloys", Phys. Rev., B19 (1979) 4896-900 (AQS)(also as Report KFKI-79-4, Hungarian Acad. Sci., Budapest, January 1979, 16pp).

2096. H.J.VIND NIELSEN, "The Influence of Cr Content on Curie Temperature, Crystallization Temperature and Room Temperature Electrical Resistivity of $Fe_{85-x}Cr_xB_{15}$ Metallic Glasses", J. Mag. Mag. Mater., 12 (1979) 187-90 (T, E, Mag).

2097. H.J.VIND NIELSEN, "A Comment on the Glass Forming Ability of Binary Alloy of Transition Metals with Metalloids", J. Non-Cryst. Solids, 33 (1979) 285-9 (AS).

2098. H.J.VIND NIELSEN, "Curie Temperature, Crystallization Temperature and Electrical Resistivity as a Function of Composition for $Fe_{80-x}Mo_xB_{20}$ Metallic Glasses", Solid State Commun., 30 (1979) 239-42 (Mag, T, E).

2099. H.J.VIND NIELSEN, "The Eutectic Compositions as a Basis for the Formation of Metallic Glasses in the Binary Alloys of Iron Group Transition Metals with Metalloids", Z. Metallkde., 70 (1979) 180-4 (AS).

2100. H.J.VIND NIELSEN, "An Estimate of the Glass Transition Temperature of Transition Metals Alloyed with Metalloids", Z. Metallkde., 70 (1979) 606-8 (T).

2101. E.N.VLASOVA and B.V.MOLOTILOV, "Sequence of Structural Changes at the Heating of Amorphous Alloys of the System Co-Fe-Si-B", Doklady Akad. Nauk. SSSR, 249 (1979) 1112 (in Russian) (AB).

2102. J.L.WALTER, "Chemical Ordering in Some Amorphous Alloys", Mater. Sci. & Engg., 39 (1979) 95-9 (AQS, SMA).

2103. R.WANG, "Short-Range Structure for Amorphous Intertransition Metal Alloys", Nature, 278 (1979) 700-4 (SMA).

2104. R.WANG, "Reply to 'A Comment on Short-Range Structure of Amorphous Alloys'", Nature, 280 (1979) 847-8 (SMA)(see also entry 1804).

2105. K.D.WARD, G.HUNGER, B.G.LEWIS and H.A.DAVIES, "The Embrittlement and Crystallization Behavior of Some Glassy Alloys", in Phase Transformations (Institution of Metallurgists, London, 1979), Part II, pp. 33-5 (M, AB)(see Phys. Abstr., 83(1980) 7685).

2106. Y.WASEDA, "A Structural Study of Amorphous Alloys by X-Ray Diffraction", Kinzoku, 49, No. 11 (1979) 16-24 (in Japanese) (R, SMA).

2107. Y.WASEDA, K.T.AUST and T.MASUMOTO, "Structural Changes in Amorphous $Pd_{77}Si_{17}Cu_6$ due to Cold Rolling and Low Temperature Annealing", Scripta Met., 13 (1979) 187-90 (AB).

2108. Y.WASEDA and T.EGAMI, "Effect of Low Temperature Annealing and Deformation on the Structure of Metallic Glasses by X-Ray Diffraction", J. Mater. Sci., 14 (1979) 1249-53 (AB).

2109. R.M.WATERSTRAT, F.HAENSSLER and J.MÜLLER, "Nb-Si A15 Compounds Produced by Liquid Quenching", J. Appl. Phys., 50 (1979) 4763-6 (MCS, S).

2110. D.C.WEBB, D.W.FORESTER, A.K.GANGULY and C.VITTORIA, "Applications of Amorphous Magnetic Layers in Surface-Acoustic-Wave Devices", IEEE Trans. Mag., MAG-15 (1979) 1961-2 (Mag, A).

2111. S.H.WHANG, L.T.KABACOFF, D.E.POLK and B.C.GIESSEN, "Measruement of Young's Modulus on Small Samples of Amorphous Metals Using the Impulse Induced Resonance Technique", Met. Trans., A10 (1979) 1789-93 (M).

2112. F.WIESER, "Differences in the Radial Distribution Functions of Metal-Metal, Metal-Metalloid and Metalloid-Metalloid Pairs in Amorphous $(Fe_{1-x}Mn_x)_{75}P_{15}C_{10}$", J. Mag. Mag. Mater., 14 (1979) 237-8 (AQS).

2113. A.R.WILLIAMS and W.L.JOHNSON, "The Structure of Some Refractory Transition Metal-Metalloid Glasses", J. Non-Cryst. Solids, 34 (1979) 121-6 (AQS).

2114. C.G.WINDSOR, H.KHEYRANDISH and M.C.NARASIMHAN, "The Vibrational Spectrum of $Pd_{80}Si_{20}$ Metal Glass", Phys. Letters, A70 (1979) 485-8 (AQS).

2115. E.D.WOHLFARTH, "Invar Behavior in Crystalline and Amorphous Alloys", J. Mag. Mag. Mater., 10 (1979) 120-5 (Mag).

2116. J.V.WOOD, "Rapid Casting of Metallic Wires", in Solidification and Casting of Metals, (The Metals Society, London, 1979) pp. 496-500 (QT).

2117. J.V.WOOD and J.V.BEE, "Rapid Solidification Processing of Nickel-Base Superalloys", in Phase Transformations, (Institution of Metallurgists, London, 1979), Part. II, pp, 25-7 (AQS).

2118. J.V.WOOD and J.V.BEE, "Precipitation and Deformation of Rapidly Solidified Nickel Superalloys", in Proc. Fifth Internat. Conf. on Strength of Metals and Alloys, eds. P.Haasen, V.Gerold and G.Kostorz (Pergamon Press, Oxford, 1979) pp.711-7 (AQS, M).

2119. J.V.WOOD and R.W.K.HONEYCOMBE, "Splat Quenching of a Ni-Cr Steel Containing B and Ti Additions", Mater. Sci. & Engg., 38 (1979) 217-26 (AQS, AB).

2120. J.V.WOOD, P.F.MILLS, J.K.BINGHAM and J.V.BEE, "Structure and Initial Precipitation in a Rapidly Solidified Nickel Superalloy", Met. Trans., A10 (1979) 575-84 (AQS, AB).

2121. J.V.WOOD and W.M.STOBBS, "Memory Alloy Wires and Ribbon Produced by Rapid Solidification", in Phase Transformations, (Institution of Metallurgists, London, 1979) Part IV, pp.20-2 (AQS)(see Phys. Abstr., 83 (1980) 7564).

2122. A.A.YAKUNIN, V.I.TKACH, A.B.LYSENKO and S.I.RYABTSEV, "Kinetic Equation of Crystallization in Continuous Cooling and Conditions for Noncrystalline Hardening", Fiz. Kinet. Tverd. Tel. Cheboksary, No. 3 (1979) 33-7 (in Russian) (PM)(see Chem. Abstr., 92(1980) 67802f).

2123. S.YAMAGUCHI, "Magnetic Analysis of an Amorphous Alloy Magnet by Electron Diffraction", J. Appl. Phys., 50 (1979) 559-60 (Mag).

2124. R.YAMAMOTO and M.DOYAMA, "The Polyhedron and Cavity Analyses of a Structural Model of Amorphous Iron", J. Phys. F : Metal Phys., 9 (1979) 617-27 (SMA).

2125. R.YAMAMOTO, H.MATSUOKA and M.DOYAMA, "A Three-Dimensional Computer Simulation for the Tensile Deformation of Amorphous Iron", Phys. Stat. Sol., (a) 51 (1979) 163-72 (M).

2126. R.YAMAMOTO, T.MIHARA, K.TAIRA and M.DOYAMA, "Amorphous Structures of Iron Obtained by Quenching of the Liquid State", Phys. Letters, A70 (1979) 41-3 (AS).

2127. H.L.YEH and R.MADDIN, "Crystallization of Amorphous Fe-P-C Alloys", Met. Trans., A10 (1979) 771-81 (AB).

2128. J.G.ZHAO, "Structural Phase Transition of Amorphous Alloys", Wuli, 8 (1979) 123-7 (in Chinese) (R)(see Chem Abstr., 91 (1979) 99988s).

2129. Z.B.ZHENG, "Formation of Amorphous Solids", Acta Met. Sinica 15 (1979) 155-66 (in Chinese) (R)(see Chem. Abstr., 91(1979) 220403c).

2130. Z.B.ZHENG, "Certain Criteria for the Formation of Amorphous Solids", Acta Met. Sinica, 15 (1979) 292-5 (in Chinese) (PM) (see Phys. Abstr., 83(1980) 19702).

2131. R.ZUBEREK and H.SZYMCZAK, "Models of Magnetostriction in Amorphous Ferrimagnetic Alloys", in 9th Internat. Colloq. Magnetic Films Surfaces, ed. L.wojtczak, Lodz. Univ., Poland, 1979) pp. 100-1 (see Chem. Abstr., 91(1979) 203290w).

AMORPHOUS MATERIALS ISSUE III (Sci. Rep. RITU, A27 1979).

2132. H.S.CHEN, "Atomic Transport Behaviour in Metallic Glasses", pp. 97-109 (T).

2133. H.FUJIMORI and N.S.KAZAMA, "Magnetic Properties and Preparation of Bulk Amorphous Alloys by High Rate Sputter Deposition", pp. 177-92 (Sp; Mag).

2134. K.FUKAMICHI, M.KIKUCHI and T.MASUMOTO, "Elastic Properties and Temperature Coefficient of Delay Time in Amorphous Pd-Si Alloys", pp. 202-9 (M, Mag).

2135. K.HASHIMOTO, K.ASAMI, M.NAKA and T.MASUMOTO, "Effect of Molybdenum on the Corrosion Behaviour of Amorphous Fe-Mo-C Alloys in 1N HCl", pp. 237-45 (C).

2136. A.INOUE and T.MASUMOTO, "Crystallization Behaviour of Amorphous High-Carbon Alloy Steels", pp. 147-58 (AB).

2137. A.INOUE, T.MASUMOTO, M.KIKUCHI and T.MINEMURA, "Effect of Compositions on the Crystallization of (Fe, Ni, Co)-Si-B Amorphous Alloys", pp. 127-46 (AB).

2138. A.INOUE, T.MASUMOTO and H.M.KIMURA, "Embrittlement of Fe, Ni, Co Amorphous Alloys Containing Metalloid Elements by Crystallization", pp. 159-71 (M, AB).

2139. N.S.KAZAMA, H.FUJIMORI and H.WATANABE, "Magnetic Properties of Bulk Amorphous La-Fe Alloys", pp. 193-201 (Mag).

2140. M.KIKUCHI, K.FUKAMAICHI and T.MASUMOTO, "Invar- and Elinvar-Type Amorphous Fe-Cr-B Alloys with High Corrosion Resistance", pp. 210-9 (M, Mag).

2141. T.MASUMOTO and H.M.KIMURA, "Delayed Fracture of Amorphous Alloys in Atmosphere", pp. 172-6 (M).

2142. M.NAKA, K.HASHIMOTO and T.MASUMOTO, "Effects of Alloying Elements on the Corrosion Behaviour of Amorphous Cobalt-base Alloys", pp. 246-58 (C).

2143. M.NAKA and T.MASUMOTO, "Effect of Metalloid Elements on the Thermal Stability of Amorphous Iron-Base Alloys", pp. 118-26 (T, AS).

2144. K.SUZUKI and T.FUKUNAGA, "A Geometrical Relaxation Model on Structure Change and Kinetics during Low Temperature Annealing of Amorphous Metals", pp. 110-7 (SMA, AB).

AMORPHOUS ALLOYS AND MATERIALS (Proceedings of the 21st Metallurgy Colloquium held at Saclay, 27-29 June 1978), C.E.N., Saclay, 1979 (All articles in French unless mentioned otherwise)).

2145. G.BELLESSA, "Low Temperature Acoustic Properties of Amorphous Materials", pp. 341-50 (PP)(see Chem.Abstr., 92(1980) 82617t).

2146. B.BOUCHER, "Magnetic Properties of Bulk Amorphous Alloys", pp. 405-21 (Mag)(see Chem.Abstr., 92(1980) 86967e).

2147. J.C.BOURGOIN, "Formation of Amorphous Layers by Irradiation", pp. 351-77 (AS)(see Chem. Abstr., 92(1980) 86035z).

2148. M.CAGNON, "Plasticity of Amorphous Alloys", pp. 255-66 (M) (see Chem Abstr., 92(1980) 98046s).

2149. R.W.CAHN, "Order and Holes in Metallic Glasses", (in English), pp. 137-57 (AQS, T).

2150. R.CASTANET and C.BERGMAN, "Thermodynamic Properties of Liquid Germanium and Tellurium Alloys", pp. 461-74 (T).

2151. DA CUNHA BELO, B.RONDOT and E.NAVARRO, "Study of the Corrosion of Amorphous Binary Alloys : Metal-Metalloid and Metal-Metal", pp. 285-99 (Sp; ED; C)(see Chem. Abstr., 92(1980) 66680c).

2152. J.DELAPLACE, "Recrystallization and Effects of Irradiation on an Amorphous Alloy $Fe_{40}Ni_{38}Mo_{4}B_{18}$", pp. 475-6 (AB).

2153. J.DIXMIER, "Techniques for Preparation of Amorphous Materials", pp. 31-58 (QT)(see Chem. Abstr., 92(1980) 82550r).

2154. P.DUWEZ, "Metallic Glasses - Their Scientific and Technical Importance", pp. 13-29 (R).

2155. A.FERT, "Electrical Transport Properties of Metallic Amorphous Materials", pp. 131-3 (E, T)(see Chem. Abstr., 92(1980) 86329y).

2156. J.FRIEDEL, "Remarks on the Atomic and Electronic Structure of Amorphous Materials", pp. 59-96 (R, AQS)(see Chem. Abstr., 92(1980) 62527k).

2157. M.GASGNIER and C.COLLIEX, "Amorphous Gadolinium-Iron Thin Films Deposited at Ambient Temperature : Preparation, Characterization and Magnetic Properties", pp. 491-501 (VD; AQS, Mag)(see Chem Abstr., 92(1980) 87006c).

2158. J.P.GASPARD, "Electronic Properties and Stability of Amorphous Alloys", pp. 111-29 (E).

2159. K.HASHIMOTO and T.MASUMOTO, "Corrosion Behavior of Amorphous Alloys", (in English) pp. 231-53 (C)(see Chem.Abstr., 92 (1980) 66614j).

2160. C.JANOT, Ph.MANGIN, G.MARCHAL and M.PIECUCH, "Principles and Methods for Characterising Amorphous Materilas", pp. 285-99 (AQS)(see Chem Abstr., 92(1980) 64856r).

2161. H.JOUVE and G.NICOLAS, "Applications of Amorphous Magnetic Alloys", pp. 437-57 (A, Mag)(see Chem. Abstr., 92(1980) 86968f).

2162. J.LIVAGE, "Amorphous Semiconductor Oxides", pp. 267-84 (AQS).

2163. F.MACHIZAUD, F.A.KUHNAST, J.FLECHON, B.AUGUIN and A.DEFRESNE, "Electronic Transport and Transformation Kinetics in Amorphous $Ni_{66.5}B_{33.5}$ and $Ni_{71}B_{29}$ Alloys", pp. 477-89 (E, AB)(see Chem. Abstr., 92(1980) 97988p).

2164. G.MARCHAL and Ph.MANGIN, "Preparation of Amorphous Fe-Si Alloys", pp. 503-22 (VD; AQS)(see Chem Abstr., 92(1980)97463p).

2165. C.MAZIERES, "Stability of Amorphous Solids", pp. 159-77 (AQS).

2166. I.B.PUCHALSKA, "Some Aspects of Magnetic Domains in Amorphous Thin Films", (in English), pp. 423-36 (Mag).

2167. J.F.SADOC, "Structure of Metallic Amorphous Alloys", pp.97-109 (AQS).

1980

2168. R.V.ALDRIDGE, "Conjectures on Mixed Conduction in an Homogeneous Sample as a Possible Explanation of the Low Temperature Resistance Minimum of Some Amorphous Metals", J. Phys. F : Metal Phys., 10 (1980) 299-308 (E).

2169. A.AMAMOU, "d Band Structure and Alloying Effects in Crystalline and Amorphous Zr-Co and Zr-Ni", Solid State Commun., 33 (1980) 1029-34 (AQS).

2170. A.AMAMOU and G.KRILL, "Photoemission and Band Structure of Crystalline and Amorphous Iron Phosphoborides", Solid State Commun., 33 (1980) 1087-90 (AQS).

2171. P.M.ANDERSON III and A.E.LORD, Jr., "Exact Values for Crystallization Temperatures at Different Heating Rates from Isothermal Transformation Times", Mater. Sci. & Engg., 43 (1980) 93-6 (T).

2172. P.M.ANDERSON III and A.E.LORD, Jr., "Viscosity of Metglas 2826 near the Glass Transition Using Rapid Heating", J. Non-Cryst. Solids, 37 (1980) 219-29 (PP).

2173. A.S.ARGON and H.Y.KUO, "Free Energy Spectra for Inelastic Deformation of Five Metallic Glassy Alloys", J. Non-Cryst. Solids, 37 (1980) 241-66 (M).

2174. K.ASO, Y.MAKINO, S.UEDAIRA, S.ITO and M.HAYAKAWA, "Observation of Magnetic Hysteresis Loop of the Perminvar Type in Worked Co-Based Amorphous Alloys", Appl. Phys. Letters, 36 (1980) 339-41 (Mag).

2175. M.BALANZAT, "A Resistometric Study of Short-Range Ordering in an Amorphous $Fe_{40}Ni_{40}P_{14}B_6$ Alloy", Scripta Met., 14 (1980) 173-8 (AB).

2176. J.BALOGH, I.DEZSI, B.FOGARASSY, L.GRANASSY, D.L.NAGY, I.VINCZE and S.ARAJS, "Influence of Atomic Substitution on Short-Range Order in Amorphous $Fe_{84}B_{16-x}C_x$ Alloys", J. Physique, 41 (1980) C1-253-4 (AQS).

2177. J.BALOGH, Gy.FAIGEL, M.TEGZE, T.KEMÈNY, A.S.SCHAAFSMA, I.VINCZE and F.VAN DER WOUDE, "Comparison of the Short-Range Order of Amorphous and Crystalline (Fe, Ni)B Alloys", J. Physique, 41 (1980) C1-255-6 (AQS).

2178. R.BASU, "Computer Study of the Composition of Some Amorphous Alloys - Metallic Glasses", Indian J. Pure & Appl. Phys., 18 (1980) 67-70 (PM).

2179. S.M.BHAGAT, M.L.SPANO, H.S.CHEN and K.V.RAO, "Magnetization of Amorphous Alloys - Deviations from Spin Wave Theory at Low Temperatures", Solid State Commun., 33 (1980) 303-7 (Mag).

2180. K.H.J.BUSCHOW, H.A.ALGRA and R.A.HENSKENS, "Magnetic Properties and Ferromagnetic Resonance in Amorphous Gd Alloys", J. Appl. Phys., 51 (1980) 561-6 (Mag).

2181. K.H.J.BUSCHOW and A.G.DIRKS, "On the Crystallization Behavior of Amorphous Alloys of Rare Earths and 3d Transition Metals", J. Phys. D : Appl. Phys., 13 (1980) 251-8 (AB, Mag).

2182. R.W.CAHN, "Metallic Glasses", Contemp. Phys., 21 (1980)43-75 (R).

2183. R.W.CAHN, J.E.EVETTS, J.PATTERSON, R.E.SOMEKH and C.K.JACKSON, "Direct Measurement by Secondary-Ion Mass Spectrometry of Self-Diffusion of Boron in $Fe_{40}Ni_{40}B_{20}$ Glass", J. Mater. Sci., 15 (1980) 702-10 (T).

2184. E.CARTIER, Y.BAER, M.LIARD and H.-J.GÜNTHERODT, "High Resolution XPS Studies of Metallic Glasses", J. Phys. F : Metal Phys., 10 (1980) L21-6 (AQS).

2185. M.CEDERGREN and G.BÄCKSTRÖM, "Pressure and Heating Rate Dependence of the Crystallization Temperature for Amorphous $(Fe_{65}Ni_{35})_{75}P_{14}B_6Al_3$ and $Ti_{50}Be_{40}Zr_{10}$", J. Non-Cryst. Solids, 37 (1980) 213-7 (T).

2186. W.CHAMBRON and A.CHAMBEROD, "Kinetics of Induced Magnetic Directional Order in a $Fe_{40}Ni_{40}P_{14}B_6$ Amorphous Alloy", Solid State Commun., 33 (1980) 157-9 (Mag).

2187. K.CHATTOPADHYAY and P.RAMACHANDRARAO, "Rapid Solidification and Decomposition of a Hypomonotectic Al-Cd Alloy", J. Mater. Sci., 15 (1980) 685-92 (AQS, AB).

2188. K.CHATTOPADHYAY, A.P.RAMINENI and P.RAMACHANDRARAO, "Estimation of Cooling Rates in Rapid Solidification from Eutectic Cell Spacing", J. Mater. Sci., 15 (1980) 797-9 (PM).

2189. P.CHAUDHARI, B.C.GIESSEN and D.TURNBULL, "Metallic Glasses", Sci. American, 242, No. 4 (1980) 98-117 (R).

2190. H.S.CHEN, "The Influence of Structure on Electrical Resistivities of Pd-Au-Si and Au-Ge-Si Glass Forming Alloys", Solid State Commun., 33 (1980) 915-9 (E).

2191. B.G.CLEMENS, W.L.JOHNSON and J.BENNETT, "Flux Pinning by Crystalline Precipitates in an Amorphous Superconducting Matrix", J. Appl. Phys., 51 (1980) 1116-20 (S).

2192. L.A.DAVIS and Y.T.YEOW, "Flow and Fracture of a Ni-Fe Metallic Glass", J. Mater. Sci., 15 (1980) 230-6 (M).

2193. M.DOI, T.HANAMURA, M.MATSUI and T.IMURA, "Electrical Resistivity and Cyclic Deformation of Liquid-Quenched Fe-Based and Co-Based Amorphous Alloys", Japan. J. Appl. Phys., 19 (1980) 449-52 (E, M).

2194. G.DUBLON, "Long-Range Interactions and Spin Wave Excitations in $Fe_xPd_{82-x}Si_{18}$ Metallic Glasses", Solid State Commun., 33 (1980) 1195-9 (Mag).

2195. G.DUBLON, C.H.LIN and J.BEVK, "The Magnetic Phase Diagram of Amorphous $Fe_xPd_{82-x}Si_{18}$, $2 \leq x \leq 25$", Phys. Letters, A76 (1980) 92-4 (Mag).

2196. J.M.DUBOIS, G.LE CAER, A.AMAMOU and U.HEROLD, "Mössbauer Study of Amorphous Fe-B, (Fe, Ni)SiB, (Fe, Ni)PB Alloys", J. Physique, 41 (1980) C1-247-8 (AQS).

2197. J.FAVARON, M.E.DE LA CRUZ, P.ESQUINAZI and F.DE LA CRUZ, "Phase Diagram and Superconducting Properties of Splat-Cooled Cu_xLa_{1-x}", Phys. Rev., B21 (1980) 2804-8 (S).

2198. V.FRANETOVIĆ, M.STUBIČAR and A.BONEFAČIĆ, "Electron Microscopic Study of Grain Size Distribution Function in Splat-Cooled Aluminium", J. Mater. Sci., 15 (1980) 353-8 (PM).

2199. R.L.FREED and J.B.VANDER SANDE, "The Metallic Glass $Cu_{56}Zr_{44}$: Devitrification and the Effects of Devitrification on Mechanical Properties", Acta Met., 28 (1980) 103-21 (AB, M).

2200. H.FUJIMORI and N.S.KAZAMA, "Sputtered Amorphous Soft Magnetic Materials", Kinzoku, 50, No. 2 (1980) 57-63 (in Japanese) (Sp; R, Mag).

2201. M.R.J.GIBBS and J.E.EVETTS, "On the Cold Rolling of Amorphous Alloys", Scripta Met., 14 (1980) 63-6 (M).

2202. U.GONSER, "Amorphous Metals and γ-Fe", J. Physique, 41 (1980) C1-51-8 (R).

2203. M.GOTO, T.TOKUNAGA, H.TANGE and T.HAMATAKE, "Magnetic Properties of the Amorphous Alloy System $(Fe_{1-x}Ni_x)_{77}Si_{10}B_{13}$", Japan. J. Appl. Phys., 19 (1980) 51-4 (Mag).

2204. V.GREGOR and J.KADLECOVÁ, "Low Temperature Resistivity of Amorphous Fe-Ni-Based Alloys", Phys. Stat. Sol., (b) 97 (1980) K49-51 (E).

2205. J.HAFNER, "Theory of Formation of Metallic Glasses", Phys. Rev., B21 (1980) 406-26 (PM).

2206. M.HAMMERSCHMIDT, H.KREYE and B.MEINCK, "On the Weldability of Metallic Alloy Glasses", Metall, 34 (1980) 23-7 (in German) (F).

2207. K.HANDRICH, "The Origin of Additional Low Energy States in Amorphous Solids", Phys. Stat. Sol., (b) 97 (1980) K37-40 (AQS).

2208. K.HANDRICH, "A New of Elementary Excitations in Amorphous Solids (Amorphons)", Phys. Stat. Sol., (b) 97 (1980) K71-5 (AQS).

2209. K.HASHIMOTO, "The Corrosion Behavior of Amorphous Alloys", Kinzoku, 50, No. 4 (1980) 42-7 (in Japanese) (R, C).

2210. M.HASHIMOTO, T.NIIZEKI and K.KAMBE, "Effect of Substrate Temperature on Crystallization of Amorphous Antimony Films", Japan. J. Appl. Phys., 19 (1980) 21-3 (AB).

2211. A.INOUE, K.KOBAYASHI, C.SURYANARAYANA and T.MASUMOTO, "An Amorphous Phase in Co-Rich Co-Ti Alloys", Scripta Met., 14 (1980) 119-23 (AS, AB, M).

2212. A.INOUE, S.SAKAI, H.M.KIMURA, T.MASUMOTO and A.HOSHI, "Superconductivity of Mo-Si-B and W-Si-B Amorphous Alloys Obtained by Liquid Quenching", Scripta Met., 14 (1980) 235-9 (S).

2213. T.IWADACHI, A.INOUE, T.MINEMURA and T.MASUMOTO, "Nonequilibrium Phases in Fe-X-C (X=Cr, Mo, W) Ternary Alloys Quenched Rapidly From Melts", J. Japan Inst. Metals, 44 (1980) 245-54 (in Japanese) (AS, MCS).

2214. J.JÄCKLE, "Effect of Electron-Phonon Interaction on the Thermopower of Amorphous Metals", J. Phys. F : Metal Phys., 10 (1980) L43-6 (VD; T).

2215. P.JACOBAEUS, J.U.MADSEN, F.KRAGH and R.M.J.COTTERILL, "Triplet Correlation in Metallic Glasses. A Field-Ion Microscopy Study", Phil. Mag., B41 (1980) 11-20 (AQS).

2216. H.J.F.JANSEN, D.S.BOUDREAUX and H.SNIJDERS, "Short-Range Order of Dense-Random Packing Models of Metallic Glasses", Phys. Rev., B21 (1980) 2274-7 (SMA).

2217. M.KAJIURA, "Effect of Chromium on the Thermal Stability of Cobalt-Based Amorphous Alloy", Japan. J. Appl. Phys., 19 (1980) 559-60 (T).

2218. Eu. E.KALININ and I.V.ZOLOTUKHIN, "New Relaxation Process Observed in Amorphous La-Al Alloys", Fiz. Tverd. Tela SSSR, 22, No. 1 (1980) 223 (in Russian) (AB).

2219. J.KÄSTNER, H.-J.SCHINK and E.F.WASSERMANN, "Electrical Resistance and Magnetoresistance of Amorphous Alloys $Gd_{67}Co_{33}$ and $Pd_{80}Si_{20}$", Solid State Commun., 33 (1980) 527-30 (E, Mag).

2220. A.KAWASHIMA, K.HASHIMOTO and T.MASUMOTO, "Fractgraphic Study of Amorphous Iron-Base Alloys Embrittled by Hydrogen and Heat Treatment", Scripta Met., 14 (1980) 41-5 (M).

2221. S.KOBAYASHI, K.MAEDA and S.TAKEUCHI, "Computer Simulation of Atomic Structure of $Cu_{57}Zr_{43}$", J. Phys. Soc. Japan, 48 (1980) 1147-52 (SMA).

2222. O.KOHMOTO, H.FUJISHIMA and T.OJIMA, "Magnetic Annealing of Zero Magnetostrictive Amorphous Alloy with High Saturation Induction", IEEE Trans. Mag., MAG-16 (1980) 440-3 (Mag).

2223. J.K.KRAUSE, T.C.LONG, T.EGAMI and D.G.ONN, "Low Temperature Specific Heat of the Metallic Glasses $Fe_xNi_{80-x}P_{14}B_6$ and $(Fe_yNi_{100-y})_{79}P_{13}B_8$ for the Spin-Glass and Spin-Cluster-Glass Regime", Phys. Rev., B21 (1980) 2886-96 (T).

2224. R.KRISHNAN, SHIVA PRASAD, J.SZTERN, C.BATTAREL and R.MORILLE, "Spin Wave Excitation Studies in Amorphous $Ni_{50}Co_{40}P_{10}$ Films", Phys. Rev., B21 (1980) 1246-9 (CD; AQS, Mag).

2225. H.J.LEAMY, J.C.BEAN, J.M.POATE and G.K.CELLER, "Nonequilibrium Incorporation of Impurities During Rapid Solidification", J. Cryst. Growth, 48 (1980) 379-82 (PM).

2226. M.LASOCKA, "Reproducibility of Thermal Effects Within the Glass Transition Region", J. Mater. Sci., 15 (1980) 1283-92 (T).

2227. F.E.LUBORSKY, J.L.WALTER and E.P.WOHLFARTH, "The Saturation Magnetization, Curie Temperature and Size Effect of Amorphous Iron Alloys", J. Phys. F : Metal Phys., 10 (1980) 959-66 (Mag).

2228. F.E.LUBORSKY and H.H.LIEBERMANN, "Stress Relaxation of Iron-Boron Amorphous Alloys", J. Appl. Phys., 51 (1980) 796 only (M).

2229. K.MASUI, M.TACHIHARA, T.YAMADA and T.TSUJIMOTO, "Phase Separation Behavior of Electrodeposited Ni-P Amorphous Alloys", J. Japan Inst. Metals, 44 (1980) 124-31 (in Japanese) (ED; AB).

2230. T.MASUMOTO, A.INOUE, S.SAKAI, H.M.KIMURA and A.HOSHI, "Superconductivity of Ductile Nb-Based Amorphous Alloys", Trans. Japan Inst. Metals, 21 (1980) 115-22 (S).

2231. M.MATSUURA, T.NOMOTO, F.ITOH and K.SUZUKI, "XPS Study of the Valance Band Spectrum of Amorphous Fe-B Alloys", Solid State Commun., 33 (1980) 895-7 (AQS).

2232. T.MATSUYAMA, K.OHTA, M.KAJIURA and T.TERANISHI, "Disaccommodation of Amorphous Magnetic Alloy $Fe_5Co_{70}Si_{10}B_{15}$", Japan. J. Appl. Phys., 19 (1980) 55-8 (Mag).

2233. T.MINEMURA, A.INOUE and T.MASUMOTO, "Nonequilibrium Austenite Phase in Rapidly Quenched Fe-Cr-C Alloys", Tetsu to Hagane, 66 (1980) 382-9 (in Japanese) (MCS).

2234. T.MIZOGUCHI, "Structural Model of Amorphous Alloys", Solid State Phys., 15 (1980) 82-5 (in Japanese) (R, SMA).

2235. A.MOGRO-CAMPERO, "Application of Extended Ziman Theory to the Temperature Dependence of the Resistivity of Amorphous Ferromagnetic Fe-B", Phys. Letters, A76 (1980) 315-7 (E).

2236. R.MOSSERI and J.DIXMIER, "Dimensionality of Subnetworks in Amorphous Covalent Alloys", J. Physique Letters, 41 (1980) 5-7 (SMA).

2237. K.NARITA, J.YAMASAKI and H.FUKUNAGA, "Measurement of Saturation Magnetostriction of a Thin Amorphous Ribbon by means of Small-Angle Magnetization Rotation", IEEE Trans. Mag., MAG-16 (1980) 435-9 (Mag).

2238. K.OHMORI, K.I.ARAI and N.TSUYA, "Ribbon-Form Sendust Alloy Made by Rapid Quenching Methods", Appl. Phys., 21 (1980) 335-8 (QT, M, Mag).

2239. D.PAVUNA, "On the Improvements of Metallic Glass Ribbons' Geometry - the Quenching Stabilizer", J. Non-Cryst. Solids, 37 (1980) 133-7 (PM).

2240. M.PIECUCH, C.JANOT and G.MARCHAL, "Charge Transfer and Atomic Volume Effect in Fe_xSn_{1-x} Amorphous Alloys", J. Physique, 41 (1980) C1-251-2 (VD; AQS).

2241. S.J.POON, "Attenuation of Magnetic Interactions in Amorphous Metals", Phys. Rev., B21 (1980) 343-6 (Mag).

2242. D.C.PRICE, S.J.CAMPBELL and P.J.BACK, "Distribution of Electric and Magnetic Hyperfine Interactions in an Amorphous Alloy", J. Physique, 41 (1980) C1-259-60 (AQS).

2243. A.K.RAYCHAUDHURI and R.HASEGAWA, "Phonon Scattering by Electrons in Noncrystalline Alloys", Phys. Rev., B21 (1980) 479-85 (AQS).

2244. B.RODMACQ, M.PIECUCH, C.JANOT, G.MARCHAL and Ph.MANGIN, "Structure and Magnetic Properties of Amorphous Fe_xSn_{1-x} Alloys", Phys. Rev., B21 (1980) 1911-23 (VD; AQS, Mag).

2245. M.P.ROSENBLUM and D.TURNBULL, "Bias Sputter Deposition of Pd-Au-Si and Pd-Si : Composition and Thermal Behavior", J. Non-Cryst. Solids, 37 (1980) 45-51 (Sp; AS, T).

2246. S.SAKAI, A.INOUE, H.M.KIMURA and T.MASUMOTO, "Superconductivity of Ni-Nb-C and Co-Nb-C Alloys Obtained by Liquid Quenching", Scripta Met., 14 (1980) 331-5 (S).

2247. M.B.SALAMON, K.V.RAO and H.S.CHEN, "Spin-Glass-Ferromagnetic Multicritical Point in Amorphous Fe-Mn Alloys", Phys. Rev. Letters, 44 (1980) 596-8 (Mag).

2248. P.J.SCHURRER and A.H.MORRISH, "Magnetic Structure of Some Amorphous Alloys", J. Physique, 41 (1980) C1-249-50 (Mag).

2249. M.G.SCOTT, "Thermal Stability and Crystallization of Metallic Glasses", Metals Tech., 7 (1980) 133-41 (R)(see also entry 1616).

2250. S.SHIBATA, T.KANAMORI, S.MITACHI and T.MANABE, "New Binary PbF_2-AlF_3 Glasses", Mater. Res. Bull., 15 (1980) 129-37 (AS, PP).

2251. K.SHIMOMURA, P.H.SHINGU and R.OZAKI, "The Morphology of α-Fe Crystals which Grow in the Amorphous $Fe_{80}(C_{1-x}B_x)_{20}$ Alloys", J. Mater. Sci., 15 (1980) 1175-82 (AB).

2252. J.-B.SUCK, H.RUDIN, H.-J.GÜNTHERODT, H.BECK, J.DAUBERT and W.GLÄSER, "Dynamical Structure Factor and Frequency Distribution of the Metallic Glass $Cu_{46}Zr_{54}$ at Room Temperature", J. Phys. C : Solid State Phys., 13 (1980) L167-72 (AQS).

2253. L.TAKÁCS, "A Systematical Analysis of Metallic Glass Mössbauer Spectra", J. Physique, 41 (1980) C1-265-6 (AQS).

2254. T.TAKAMORI, T.MIZOGUCHI and T.R.McGUIRE, "Viscous Flow and Crystallization of a Metallic Glass", Mater. Res. Bull., 15 (1980) 81-7 (PP, AB).

2255. S.TAKAYAMA, "Studies in the Cold Drawing of $Pd_{77.5}Cu_6Si_{16.5}$ Metallic Glass and 316 Stainless Steel Crystalline Wires", J. Mater. Sci., 15 (1980) 649-56 (F).

2256. S.TAKAYAMA and R.G.FEHRMAN, "Composite Metallic Glass Wires", J. Mater. Sci., 15 (1980) 532-4 (F).

2257. S.TAKAYAMA and T.OI, "The Effects of Casting Conditions on Magnetic Properties of Amorphous Alloys", J. Japan Inst. Metals, 44 (1980) 1-6 (in Japanese) (PM, Mag).

2258. S.TAKAYAMA and T.OI, "Casting Condition of Metallic Glasses", J. Japan Inst. Metals, 44 (1980) 113-7 (in Japanese) (PM).

2259. A.I.TAUB, "Stress-Strain Rate Dependence of Homogeneous Flow in Metallic Glasses", Acta Met., 28 (1980) 633-7 (M).

2260. M.TENHOVER, "Superconducting Properties of Rapidly Quenched and Heat Treated Zr-V and Hf-V Foils", Appl. Phys., 21 (1980) 279-82 (S).

2261. K.TOGANO, H.KUMAKURA and K.TACHIKAWA, "Metastable A15 Phase in Liquid-Quenched Nb-Si Binary Alloys", Phys. Letters, A76 (1980) 83-5 (MCS).

2262. A.M.TONEJC and A.BONEFAČIĆ, "X-Ray Line Broadening Study of Splat-Quenched Ag-Sn Alloys", J. Mater. Sci., 15 (1980) 415-25 (AQS).

2263. N.TSUYA and K.I.ARAI, "Magnetostriction of Amorphous Thin Bands and Their Applications", Kinzoku, 50, No. 3 (1980) 57-61 (in Japanese) (R, Mag).

2264. N.TSUYA, K.I.ARAI, T.TAKEUCHI, K.OHMORI, T.OJIMA and A.KUROIWA "Silicon Ribbons Made by Roller Quenching Methods", J. Electron Mater., 9 (1980) 111-28 (QT, AQS).

2265. F.VAN DER WOUDE and I.VINCZE, "Magnetism from Mössbauer Spectroscopy", J. Physique, 41 (1980) C1-151-4 (R, Mag).

2266. I.VINCZE, T.KEMÈNY and S.ARAJS, "Short-Range Order in Transition Metal-Metalloid Glasses", Phys. Rev., B21 (1980) 937-40 (AQS).

2267. I.VINCZE, F.VAN DER WOUDE and J.BALOGH, "Short-Range Order in Transition Metal-Metalloid Glasses", J. Physique, 41 (1980) C1-257-8 (AQS).

2268. B.J.WACLAWSKI and D.S.BOUDREAUX, "Ultraviolet Photoelectron Spectroscopy of Pd-Si Glasses", Solid State Commun., 33 (1980) 589-91 (AQS).

2269. Y.WASEDA, "The Structure of Metallic Glasses", in The Structure of Non-Crystlline Materials by Y.Waseda (McGraw-Hill, New York, 1980), Chapter 4, pp. 87-132 (R).

2270. Y.WASEDA, K.T.AUST and J.L.WALTER, "Structural Investigation of Amorphous Fe-Ni-B Alloys", J. Mater. Sci., 15 (1980) 1252-8 (AQS).

2271. T.WATANABE and M.G.SCOTT, "The Crystallization of the Amorphous Alloy $Fe_{40}Ni_{40}P_{14}B_6$", J. Mater. Sci., 15 (1980) 1131-9 (AB).

2272. G.WEISS, W.ARNOLD, K.DRANSFELD and H.-J.GÜNTHERODT, "Ultrasonic Properties of Superconducting Amorphous PdZr", Solid State Commun., 33 (1980) 111-4 (PP).

2273. S.H.WHANG, L.T.KABACOFF, D.E.POLK and B.C.GIESSEN, "A Technique for the Measurement of Young's Modulus of Small Metallic Glass Samples", J. Mater. Sci., 15 (1980) 247-8 (M).

2274. E.WIESER, "Magnetic Ordering in Amorphous $(Fe_{1-x}Mn_x)_{75}P_{15}C_{10}$ Alloys", J. Physique, 41 (1980) C1-259-60 (Mag).

2275. J.WROŃSKI, H.MATYJA, J.PIEKOSZEWSKI and J.SUWALSKI, "Mössbauer Effect Study of Magnetization Orientation in $Fe_{80}B_{20}$ Glass", Solid State Commun., 33 (1980) 1155-7 (Mag).

2276. R.YAMAMOTO and M.DOYAMA, "Computer Simulation of the Structure and Physical Properties of Amorphous Metals", Kinzoku, 50, No. 2 (1980) 64-9 (in Japanese) (R, SMA).

2277. L.ZAUSKI, A.ZAUSKA and M.LASOCKA, "Thermal Evidence of Structural Phase Transformation in Te-Ge-Sb and Te-Ge-Bi Systems", J. Mater. Sci., 15 (1980) 1331-3 (T).

AMORPHOUS MATERIALS ISSUE 1V (Sci. Rep. RITU, A28 (1980))

2278. T.FUKUNAGA and K.SUZUKI, "An Experimental Derivation of Partial Structure Functions of Amorphous $Pd_{80}Si_{20}$ Alloy Using Combination of X-ray, Electron and Neutron Diffraction Experiments", pp. 208-17 (AQS).

2279. N.HAYASHI, Y.SYONO, Y.NAKAGAWA, K.OKAMURA and S.YAJIMA, "Field-Cooling Effect in the Amorphous Magnetic Oxide", pp. 164-71 (Mag).

2280. T.HONDA, F.ITOH, T.FUKUNAGA and K.SUZUKI, "Electronic Structures of Amorphous Co-P and Ni-P Alloys by γ-Ray Compton Scattering", pp. 218-31 (AQS).

2281. A.INOUE, K.KOBAYASHI and T.MASUMOTO, "Mechanical Properties and Thermal Stability of (Fe, Co, Ni)-Mo-C Quaternary Amorphous Alloys", pp. 172-81 (M, T).

2282. A.INOUE, C.SURYANARAYANA, T.MASUMOTO and A.HOSHI, "Superconductivity of Ti-Nb-Si Alloys Crystallized from the Amorphous State", pp. 182-94 (S, AB).

2283. M.KIKUCHI, K.FUKAMICHI and T.MASUMOTO, "The Pole Effect of Ferromagnetic Thin Sheets and Stiffening of Fe-B Base Amorphous Alloys", pp. 242-54 (M).

2284. M.NAKA, K.HASHIMOTO and T.MASUMOTO, "Corrosion Behavior of Amorphous Ni-Ti-P and Ni-Cr-P Alloys", pp. 156-63 (C).

2285. M.NOSE and T.MASUMOTO, "Amorphous Phase Formation in the Zirconium-Poor Corner of (Fe, Co, Ni)-Zr Systems", pp. 232-41 (AS, T, M).

2286. K.SHIRAKAWA, Y.KOBAYASHI and T.MASUMOTO, "Crystallization and Electrical Properties of Amorphous Cu-Ag-P Alloys", pp. 255-65 (AB, E).

2287. C.SURYANARAYANA, A.INOUE and T.MASUMOTO, "An Electron Microscopic Study on the Crystallization of Amorphous Ti-(Fe, Co or Ni)-B Alloys", pp. 195-207 (AS, AB).

2288. Y.WASEDA and H.S.CHEN, "On the Structure of Metallic Glasses of Transition Metal-Metalloid Systems", pp. 143-55 (AQS).

STRUCTURE AND PROPERTIES OF AMORPHOUS METALS II, eds. T.Masumoto and T.Imura, Suppl. Sci. Rep. RITU, A28 (1980).

2289. K.I.ARAI and N.TSUYA, "Magnetomechanical Coupling Factor in Amorphous $Fe_{81}(Si, B, C)_{19}$ Ribbons", pp. 247-50 (M, Mag).

2290. K.ASO, Y.MAKINO, S.UEDAIRA, S.ITO and H.MAYAKAWA, "Effects of Cr-Substitution on Magnetic Properties in Fe-Based Amorphous Alloys", pp. 282-6 (Mag).

2291. K.DOI, "Structural Changes of a-$Pd_{80}Si_{20}$ and a-SiO_2 by Neutron Irradiations, with a Hint for the Characterization of Amorphous Materials", pp. 124-35 (AQS).

2292. F.E.FUJITA, "On the Structure and Structural Units of Amorphous Metals", pp. 1-7 (SMA).

2293. T.HANAMURA, M.DOI, M.MATSUI and T.IMURA, "Correlation between the Changes in Structure and Electrical Resistivity of Amorphous Metals", pp. 93-103 (AB, E).

2294. K.HASHIMOTO, "Characteristics of Corrosion Resistant Amorphous Alloys", pp. 201-16 (C).

2295. A.INOUE and T.MASUMOTO, "Melt-Quenched Amorphous Superconducting Alloys", pp. 165-200 (S, R).

2296. M.KIRITANI, "Variable Latent Vacancy Mechanism of Diffusion in Amorphous Metals", pp. 56-64 (T).

2297. S.KOBAYASHI, K.MAEDA and S.TAKEUCHI, "Simulation of Structure and Mechanical Properties of Amorphous $Cu_{57}Zr_{43}$", pp. 155-64 (AQS, M).

2298. H.KOMIYAMA, A.YOKOYAMA, H.INOUE, T.MASUMOTO and H.M.KIMURA, "Catalytic Properties of Amorphous $Fe_{40}Ni_{40}P_{16}B_4$ Ribbons", pp. 217-21 (C).

2299. K.MAEDA, K.MUKASA and Y.KUDO, "Crystallization of Amorphous Pd-Si Alloy during Low Temperature Annealing", pp. 118-23 (AB).

2300. T.MIHARA, S.OTAKE, H.FUKUSHIMA, M.NAKAMURA and M.DOYAMA, "Structural Relaxation in Amorphous $Fe_{40}Ni_{40}P_{14}B_6$ Alloy Studied by Positron Annihilation and Electrical Resistivity Measurement", pp. 38-45 (AB).

2301. T.MIYAZAKI and M.TAKAHASHI, "Instability of Permeability of an Amorphous $Fe_5Co_{70}Si_{15}B_{10}$ Alloy", pp. 311-9 (Mag).

2302. T.MIZOGUCHI, "Ferrimagnetism in Amorphous Alloys", pp. 237-46 (Mag).

2303. M.NAGUMO and T.SATO, "Glass-Forming Ability of Fe-Base Alloys", pp. 136-42 (PM).

2304. S.NANAO, K.TOKUMITSU, A.MATSUZAKI, H.INO and S.NISHIKAWA, "Structure and Aging Behavior of Amorphous Pd-Si Alloys", pp. 104-17 (AQS, AB).

2305. K.NARITA, H.FUKUNAGA, J.YAMASAKI and K.HARA, "Dependence of Magnetic Properties on Metalloid Content in Amorphous Fe-Si-B Alloys", pp. 251-63 (Mag).

2306. M.NOSE and T.MASUMOTO, "Characteristics of (Fe, Co and/or Ni)-Zr Amorphous Alloys", pp. 222-37 (Mag, M, T).

2307. K.OSAMURA, K.SHIBUE, R.SUZUKI and Y.MURAKAMI, "SAXS Study on the Structure and Crystallization of Amorphous Metallic Alloys", pp. 65-75 (AQS, AB).

2308. R.OSHIMA, "Short-Range Atomic Arrangement in Amorphous Fe-B Alloys", pp. 8-11 (AQS).

2309. H.SENNO, H.SAKAKIMA and E.HIROTA, "Effect of Mn-substitution on Magnetic Properties of Co-Si-B and Co-Fe-Si-B Amorphous Alloys", pp. 276-81 (Mag).

2310. P.H.SHINGU, M.CHUDO and R.OZAKI, "Dimensional Change of Amorphous Alloys due to Crystallization", pp. 85-92 (PP).

2311. P.H.SHINGU, Y.KOMATSU and R.OZAKI, "Kinetics of Crystallization of Amorphous Alloys", pp. 76-84 (AB).

2312. C.SURYANARAYANA, "Liquid-Quenched Metal-Metal Glasses", pp. 143-54 (R).

2313. K.SUZUKI, F.ITOH, M.MISAWA, M.MATSUURA, T.FUKUNAGA and K.IKENO, "Characteristic Behaviors on Atomic Structure and Electronic State of Fe-B Glassy Alloys", pp. 12-28 (AQS).

2314. Y.SYONO and Y.NAKAGAWA, "Speromagnetism in Homogeneous Oxide Glasses Containing a Large Amount of Fe^{3+} Ions", pp. 264-75 (Mag).

2315. M.TAKAHASHI, S.ISHIO and F.SATO, "Magnetism of Amorphous and Liquid States of Rapidly Quenched Metallic Alloys", pp. 287-98 (Mag).

2316. M.TAKAHASHI, S.ISHIO and T.MIYAZAKI, "Crystallization Mechanism of Amorphous Alloys - Magnetic Anistropy", pp. 299-310

2317. S.TSUKAHARA and T.TSUSHIMA, "Electron Diffraction Patterns of Gd-Fe Thin Film Deposited at Low Temperature", pp. 46-55 (VD; AQS).

2318. R.YAMAMOTO, M.NAKAMURA and M.DOYAMA, "Computer Simulations of the Structure and the Vibrational Properties of Amorphous Iron", pp. 29-37 (AQS).

J. Mag. Mag. Mater., 15-18 (1980) (Proceedings of the Internat. Conf. on Magnetism (ICM 79) held at Munchen, W.Germany).

2319. J.A.ABOAF and E.KLOKHOLM, "Magnetic Properties of Metal-Metalloid Thin Film Glasses", pp. 1385-6 (Sp; Mag).

2320. H.A.ALGRA, K.H.J.BUSCHOW and R.A.HENSKENS, "Magnetic Properties and Ferromagnetic Resonance in Amorphous Gd Alloys", pp. 1395-6 (Mag).

2321. P.ALLIA, P.MAZZETTI, G.P.SOARDO and F.VINAI, "Structural Instabilities and Magnetic Relaxation in Amorphous Ferromagnets", pp. 1361-3 (Mag).

2322. K.I.ARAI, N.TSUYA, K.OHMORI, H.SHIMANAKA and T.MIYAZAKI, "Rapidly Quenched Ribbon Form Silicon-Iron Alloy with High Silicon Concentration", pp. 1425-6 (Mag),

2323. C.AROCA, J.M.RIVEIRO, G.RIVERO and M.C.SÁNCHEZ, "Domain Structure in Some Amorphous Alloys", pp. 1379-80 (ED; Mag).

2324. A.R.BHATTI. P.J.GRUNDY, G.A.JONES and R.S.TEBBLE, "An Electron Microscope Study of Microstructural and Related Domain Changes in Annealed Fe-B Amorphous Alloys", pp. 1413-4 (AQS, Mag).

2325. S.BJARMAN, R.KAMAL and R.WAPPLING , "Mössbauer Studies of Amorphous Iron and Metglas 2826MB", pp. 1389-90 (AQS, Mag).

2326. P.BUTVIN, P.DUHAJ and Z.FRAIT, "Structural and Magnetic Heterogeneity of Ternary Amorphous Metallic Alloys", pp.1404-6 (AQS, Mag).

2327. G.DIETZ and R.DISSELNKÖTTER, "Magnetic Domains in Amorphous Fe-P Alloys", pp. 1381-2 (ED; Mag).

2328. J.DURAND, B.LEMIUS, R.HASEGAWA, D.ALIAGA-GUERRA and P.PANISSOD, "Hyperfine Field Distribuion at ^{59}Co in Glassy and Crystalline Co_3B", pp. 1373-4 (Mag).

2329. U.ERNST, W.FELSCH and K.SAMWER, "Magnetic Ordering and Crystal Field Effects in the Metallic Glass $Ce_{80}Au_{20}$", pp. 1375-6 (Mag).

2330. H.FRANKE, S.DEY, M.ROSENBERG, F.E.LUBORSKY and J.L.WALTER, "Hyperfine Fields and Local Magnetic Moments of Metallic Glasses of 3d-Transition Metals", pp. 1364-6 (Mag).

2331. U.GONSER, H.FISCHER, M.GHAFARI, H.-G.WAGNER and R.S.PRESTON, "Structure of Ferromagnetic Amorphous Metals ($M_{80}Au_{20}$ Type)", pp. 1355-6 (AQS, Mag).

2332. C.D.GRAHAM, Jr. and T.EGAMI, "Magnetic Properties of Amorphous Alloys", pp. 1325-30 (R, Mag).

2333. H.GRIMM and H.KRONMÜLLER, "Magnetic Polarization of Amorphous Alloys in the Approach to Ferromagnetic Saturation", pp.1411-2 (Mag).

2334. N.HASSANAIN, A.BERRADA, J.DURAND and B.LOEGEL, "Magnetic and Superconducting Properties of Dilute Alloys of Pr in Amorphous La_4Au", pp. 1377-8 (Mag, S).

2335. J.J.HAUSER, "Concentrated Spin Glass Systems : Amorphous MnSi and MnGe", pp. 1387-8 (VD; Mag).

2336. A.HERNANDO, J.M.BARANDIARAN, V.MADURGA, M.VAZQUEZ and E.ASCASIBAR, "The Magnetization Process in Twisted Amorphous Ribbons", pp. 1537-8 (Mag).

2337. H.R.HILZINGER and W.KUNZ, "Magnetic Properties of Amorphous Alloys with Low Magnetostriction", pp. 1357-8 (Mag).

2338. T.JAGIELINSKI, T.WALECKI, K.I.ARAI and N.TSUYA, "The SMFMR in Amorphous Fe-Co Alloys", pp. 1415-6 (Mag).

2339. J.KAMARAD, Z.ARNOLD, J.SCHNEIDER and S.KRUPICKA, "Effect of Pressure on the Curie Temperature of Metallic Glasses $(Ni_x-Fe_{1-x})_{80}P_{10}B_{10}$", pp. 1409-10 (Mag).

2340. T.KANEYOSHI, "Spin Waves in Amorphous Ferro- and Aspero-Magnets", pp. 1343-4 (Mag).

2341. T.KANEYOSHI and S.IWABUCHI, "Indirect Exchange Interaction and Paramagnetic Curie Temperatures in Amorphous Metallic Alloys", pp. 1421-2 (Mag).

2342. N.S.KAZAMA, H.FUJIMORI and H.WATANABE, "Magnetic Properties of Amorphous La-Fe Alloys by High-Rate Sputtering", pp. 1423-4 (Sp; Mag).

2343. N.S.KAZAMA, T.MASUMOTO and M.MITERA, "Effects of Metalloids on the Magnetic Properties of Iron- and Cobalt-Based Amorphous Alloys", pp. 1311-5 (R, Mag).

2344. É.KISDI-KOSZÓ, L.POTOCKÝ and A.NOVÁK, "Development of the Induced Anisotropy by Heat Treatment in a Magnetic Field and Under Applied Mechanical Stress in Amorphous Fe-B Alloys", pp. 1383-4 (Mag).

2345. F.E.LUBORSKY, J.L.WALTER, H.H.LIEBERMANN and E.P.WOHLFARTH, "The Effect of Temperature on Magnetic Saturation of Some Amorphous Alloys", pp. 1351-4 (Mag).

2346. H.MORITA, H.FUJIMORI and Y.OBI, "Roll Magnetic Anisotropy of Amorphous Alloys", pp. 1359-60 (Mag).

2347. M.MÜLLER and H.-J.GÜNTHERODT, "Magnetic Properties of Glassy and Liquid Metals", pp. 1349-50 (Mag).

2348. N.RIVIER and M.A.CONTINENTINO, "Hidden Excitations in Amorphous Ferromagnets", pp. 1419-20 (Mag).

2349. P.J.SCHURRER and A.H.MORRISH, "Mössbauer Study of Magnetic Anisotorpy in Amorphous $Fe_{40}Ni_{38}Mo_4B_{18}$ (Metglas 2826MB)", pp. 577-8 (Mag).

2350. T.SHIGEMATSU, T.SHINJO, Y.BANDO and T.TAKADA, "Mössbauer Effect of Amorphous Fe-Sb Thin Films", pp. 1367-8 (VD; Mag).

2351. A.M.STEWART, "Magnetic Resonance at 20GHz of an Amorphous Metal", pp. 1417-8 (Mag).

2352. M.TAKAHASHI and F.SATO, "Magnetic Structure in Amorphous (Fe, Co, Ni)-B Alloys", pp. 1407-8 (Mag).

2353. I.VINCZE, F.VAN DER WOUDE, T.KEMÈNY and A.S.SCHAAFSMA, "Magnetic Properties of Amorphous Transition Metal Alloys", pp. 1336-8 (Mag).

2354. D.WAGNER and E.P.WOHLFARTH, "Theory and Applications of the Landau-Ginzburg Theory of Amorphous Ferromagnetism", pp.1345-6 (Mag).

2355. K.ZÁVĚTA, J.SCHNEIDER, B.SPRINGMANN, E.KRATOCHVÍLOVÁ and S.KRUPIČKA, "Induced Anisotropy and Time Changes of Permeability in $(CoFe)_{80}B_{20}$ Metallic Glasses", pp. 1427-9 (Mag).

AUTHOR INDEX

A

ALLOY INDEX

SUBJECT INDEX

A

B

C

M

COMPOSITIONS OF COMMERCIAL AMORPHOUS ALLOYS

METGLAS 2204	$Ti_{50}Be_{40}Zr_{10}$
METGLAS 2605	$Fe_{80}B_{20}$
METGLAS 2605A	$Fe_{78}Mo_{2}B_{20}$
METGLAS 2605S	$Fe_{82}B_{12}Si_{6}$
METGLAS 2605SC	$Fe_{81}B_{13.5}Si_{3.5}C_{2}$
METGLAS 2615	$Fe_{80}P_{16}C_{3}B_{1}$
METGLAS 2826	$Fe_{40}Ni_{40}P_{14}B_{6}$
METGLAS 2826A	$Ni_{36}Fe_{32}Cr_{14}P_{12}B_{6}$
METGLAS 2826B	$Ni_{49}Fe_{29}P_{14}B_{6}Si_{2}$
METGLAS 2826MB	$Fe_{40}Ni_{38}Mo_{4}B_{18}$
VITROVAC 0040	$Fe_{40}Ni_{40}B_{20}$
VITROVAC 6010	$Co_{50}Ni_{20}Fe_{6}Si_{12}B_{12}$

GPSR Compliance

The European Union's (EU) General Product Safety Regulation (GPSR) is a set of rules that requires consumer products to be safe and our obligations to ensure this.

If you have any concerns about our products, you can contact us on ProductSafety@springernature.com

In case Publisher is established outside the EU, the EU authorized representative is:

Springer Nature Customer Service Center GmbH
Europaplatz 3
69115 Heidelberg, Germany

Batch number: 10620881

Printed by Printforce, the Netherlands